MULTIPLICATION
FACTS 0-12

with ANSWERS

$$
\begin{array}{r}
6 \\
\times 4 \\
\hline
24
\end{array}
\qquad
\begin{array}{r}
8 \\
\times 7 \\
\hline
56
\end{array}
\qquad
\begin{array}{r}
12 \\
\times 5 \\
\hline
60
\end{array}
$$

Chris McMullen, Ph.D.

Multiplication Facts 0-12 with Answers
Improve Your Math Fluency Worksheets
Chris McMullen, Ph.D.

Zishka Publishing
ISBN: 978-1-941691-52-6

Textbooks > Math > Arithmetic
Study Guides > Workbooks> Math
Education > Math > Arithmetic

CONTENTS

INTRODUCTION

Practicing multiplication facts is essential for developing fluency in mathematics. This workbook presents the facts in stages.

- The first exercise sets focus on a single number to help students become familiar with one set at a time.
- The next set eases students into it by only including 0 thru 5. This helps to build confidence.
- When 4-9 are introduced, they are paired with smaller numbers (0-5) so that the practice remains focused.
- One section is dedicated to 4-9 times 4-9. This helps to provide extra practice with the harder facts, yet also helps students learn them by focusing on them.
- A mixed section (0-9) challenges students to know all of their facts at once. There is also a section with 2-9 that removes the easiest facts (0 and 1).
- Similarly, the numbers 10-12 are introduced in stages before moving onto mixed practice.
- The book concludes with sections of 0-12 and 2-12.
- There is a handy answer key at the back of the book.

May you (or your students) find this workbook useful and become more fluent with multiplication.

Multiplication Table

	0	1	2	3	4	5	6	7	8	9	10	11	12
0	0	0	0	0	0	0	0	0	0	0	0	0	0
1	0	1	2	3	4	5	6	7	8	9	10	11	12
2	0	2	4	6	8	10	12	14	16	18	20	22	24
3	0	3	6	9	12	15	18	21	24	27	30	33	36
4	0	4	8	12	16	20	24	28	32	36	40	44	48
5	0	5	10	15	20	25	30	35	40	45	50	55	60
6	0	6	12	18	24	30	36	42	48	54	60	66	72
7	0	7	14	21	28	35	42	49	56	63	70	77	84
8	0	8	16	24	32	40	48	56	64	72	80	88	96
9	0	9	18	27	36	45	54	63	72	81	90	99	108
10	0	10	20	30	40	50	60	70	80	90	100	110	120
11	0	11	22	33	44	55	66	77	88	99	110	121	132
12	0	12	24	36	48	60	72	84	96	108	120	132	144

Visual Example 1

7×4 can be drawn as 7 groups of 4.

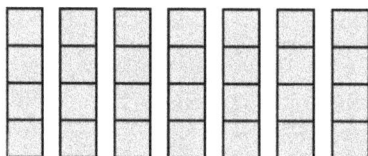

7×4 is the same as adding seven fours together.
$$7 \times 4 = 4 + 4 + 4 + 4 + 4 + 4 + 4 = 28$$

4×7 can be drawn as 4 groups of 7.

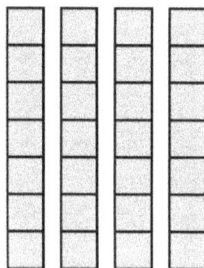

4×7 is the same as adding four sevens together.
$$4 \times 7 = 7 + 7 + 7 + 7 = 28$$

Note that order doesn't matter in multiplication.
$$7 \times 4 = 4 \times 7 = 28$$

Visual Example 2

3×6 can be drawn as **3** groups of **6**.

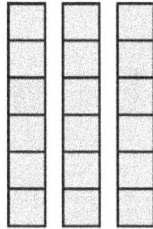

3×6 is the same as adding three sixes together.
$$3 \times 6 = 6 + 6 + 6 = 18$$

6×3 can be drawn as **6** groups of **3**.

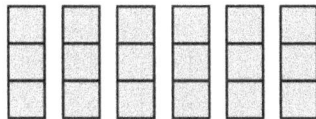

6×3 is the same as adding six threes together.
$$6 \times 3 = 3 + 3 + 3 + 3 + 3 + 3 = 18$$

Note that order doesn't matter in multiplication.
$$3 \times 6 = 6 \times 3 = 18$$

Table Example 1

	0	1	2	3	4	5	6	7	8	9	10	11	12
0	0	0	0	0	0	0	0	0	0	0	0	0	0
1	0	1	2	3	4	5	6	7	8	9	10	11	12
2	0	2	4	6	8	10	12	14	16	18	20	22	24
3	0	3	6	9	12	15	18	21	24	27	30	33	36
4	0	4	8	12	16	20	24	28	32	36	40	44	48
5	0	5	10	15	20	25	30	35	40	45	50	55	60
6	0	6	12	18	24	30	36	42	48	54	60	66	72
7	0	7	14	21	28	35	42	49	56	63	70	77	84
8	0	8	16	24	32	40	48	56	64	72	80	88	96
9	0	9	18	27	36	45	54	63	72	81	90	99	108
10	0	10	20	30	40	50	60	70	80	90	100	110	120
11	0	11	22	33	44	55	66	77	88	99	110	121	132
12	0	12	24	36	48	60	72	84	96	108	120	132	144

$$7 \times 4 = 28$$

Table Example 2

	0	1	2	3	4	5	6	7	8	9	10	11	12
0	0	0	0	0	0	0	0	0	0	0	0	0	0
1	0	1	2	3	4	5	6	7	8	9	10	11	12
2	0	2	4	6	8	10	12	14	16	18	20	22	24
3	0	3	6	9	12	15	18	21	24	27	30	33	36
4	0	4	8	12	16	20	24	28	32	36	40	44	48
5	0	5	10	15	20	25	30	35	40	45	50	55	60
6	0	6	12	18	24	30	36	42	48	54	60	66	72
7	0	7	14	21	28	35	42	49	56	63	70	77	84
8	0	8	16	24	32	40	48	56	64	72	80	88	96
9	0	9	18	27	36	45	54	63	72	81	90	99	108
10	0	10	20	30	40	50	60	70	80	90	100	110	120
11	0	11	22	33	44	55	66	77	88	99	110	121	132
12	0	12	24	36	48	60	72	84	96	108	120	132	144

$$3 \times 6 = 18$$

Study the 0 Facts

$$\begin{array}{r} 0 \\ \times\ 0 \\ \hline 0 \end{array} \quad \begin{array}{r} 0 \\ \times\ 1 \\ \hline 0 \end{array} \quad \begin{array}{r} 0 \\ \times\ 2 \\ \hline 0 \end{array} \quad \begin{array}{r} 0 \\ \times\ 3 \\ \hline 0 \end{array} \quad \begin{array}{r} 0 \\ \times\ 4 \\ \hline 0 \end{array}$$

$$\begin{array}{r} 0 \\ \times\ 5 \\ \hline 0 \end{array} \quad \begin{array}{r} 0 \\ \times\ 6 \\ \hline 0 \end{array} \quad \begin{array}{r} 0 \\ \times\ 7 \\ \hline 0 \end{array} \quad \begin{array}{r} 0 \\ \times\ 8 \\ \hline 0 \end{array} \quad \begin{array}{r} 0 \\ \times\ 9 \\ \hline 0 \end{array}$$

$$\begin{array}{r} 0 \\ \times\ 0 \\ \hline 0 \end{array} \quad \begin{array}{r} 1 \\ \times\ 0 \\ \hline 0 \end{array} \quad \begin{array}{r} 2 \\ \times\ 0 \\ \hline 0 \end{array} \quad \begin{array}{r} 3 \\ \times\ 0 \\ \hline 0 \end{array} \quad \begin{array}{r} 4 \\ \times\ 0 \\ \hline 0 \end{array}$$

$$\begin{array}{r} 5 \\ \times\ 0 \\ \hline 0 \end{array} \quad \begin{array}{r} 6 \\ \times\ 0 \\ \hline 0 \end{array} \quad \begin{array}{r} 7 \\ \times\ 0 \\ \hline 0 \end{array} \quad \begin{array}{r} 8 \\ \times\ 0 \\ \hline 0 \end{array} \quad \begin{array}{r} 9 \\ \times\ 0 \\ \hline 0 \end{array}$$

Practice with 0

① 0 × 4 ② 6 × 0 ③ 0 × 3 ④ 7 × 0 ⑤ 0 × 1 ⑥ 2 × 0 ⑦ 0 × 8 ⑧ 0 × 0 ⑨ 0 × 5 ⑩ 9 × 0

⑪ 0 × 1 ⑫ 8 × 0 ⑬ 0 × 2 ⑭ 0 × 0 ⑮ 0 × 6 ⑯ 5 × 0 ⑰ 0 × 9 ⑱ 4 × 0 ⑲ 0 × 7 ⑳ 3 × 0

㉑ 0 × 0 ㉒ 7 × 0 ㉓ 0 × 9 ㉔ 8 × 0 ㉕ 0 × 4 ㉖ 3 × 0 ㉗ 0 × 5 ㉘ 1 × 0 ㉙ 0 × 2 ㉚ 6 × 0

㉛ 0 × 6 ㉜ 4 × 0 ㉝ 0 × 5 ㉞ 9 × 0 ㉟ 0 × 2 ㊱ 1 × 0 ㊲ 0 × 0 ㊳ 7 × 0 ㊴ 0 × 3 ㊵ 8 × 0

㊶ 0 × 7 ㊷ 3 × 0 ㊸ 0 × 8 ㊹ 1 × 0 ㊺ 0 × 9 ㊻ 0 × 0 ㊼ 0 × 6 ㊽ 2 × 0 ㊾ 0 × 4 ㊿ 5 × 0

Study the 1 Facts

$$
\begin{array}{r} 1 \\ \times\ 0 \\ \hline 0 \end{array}
\qquad
\begin{array}{r} 1 \\ \times\ 1 \\ \hline 1 \end{array}
\qquad
\begin{array}{r} 1 \\ \times\ 2 \\ \hline 2 \end{array}
\qquad
\begin{array}{r} 1 \\ \times\ 3 \\ \hline 3 \end{array}
\qquad
\begin{array}{r} 1 \\ \times\ 4 \\ \hline 4 \end{array}
$$

$$
\begin{array}{r} 1 \\ \times\ 5 \\ \hline 5 \end{array}
\qquad
\begin{array}{r} 1 \\ \times\ 6 \\ \hline 6 \end{array}
\qquad
\begin{array}{r} 1 \\ \times\ 7 \\ \hline 7 \end{array}
\qquad
\begin{array}{r} 1 \\ \times\ 8 \\ \hline 8 \end{array}
\qquad
\begin{array}{r} 1 \\ \times\ 9 \\ \hline 9 \end{array}
$$

$$
\begin{array}{r} 0 \\ \times\ 1 \\ \hline 0 \end{array}
\qquad
\begin{array}{r} 1 \\ \times\ 1 \\ \hline 1 \end{array}
\qquad
\begin{array}{r} 2 \\ \times\ 1 \\ \hline 2 \end{array}
\qquad
\begin{array}{r} 3 \\ \times\ 1 \\ \hline 3 \end{array}
\qquad
\begin{array}{r} 4 \\ \times\ 1 \\ \hline 4 \end{array}
$$

$$
\begin{array}{r} 5 \\ \times\ 1 \\ \hline 5 \end{array}
\qquad
\begin{array}{r} 6 \\ \times\ 1 \\ \hline 6 \end{array}
\qquad
\begin{array}{r} 7 \\ \times\ 1 \\ \hline 7 \end{array}
\qquad
\begin{array}{r} 8 \\ \times\ 1 \\ \hline 8 \end{array}
\qquad
\begin{array}{r} 9 \\ \times\ 1 \\ \hline 9 \end{array}
$$

Practice with 1

① 1 × 7   ② 4 × 1   ③ 1 × 3   ④ 5 × 1   ⑤ 1 × 0   ⑥ 2 × 1   ⑦ 1 × 6   ⑧ 8 × 1   ⑨ 1 × 1   ⑩ 9 × 1

⑪ 1 × 0   ⑫ 6 × 1   ⑬ 1 × 2   ⑭ 8 × 1   ⑮ 1 × 4   ⑯ 1 × 1   ⑰ 1 × 9   ⑱ 7 × 1   ⑲ 1 × 5   ⑳ 3 × 1

㉑ 1 × 8   ㉒ 5 × 1   ㉓ 1 × 9   ㉔ 6 × 1   ㉕ 1 × 7   ㉖ 3 × 1   ㉗ 1 × 1   ㉘ 0 × 1   ㉙ 1 × 2   ㉚ 4 × 1

㉛ 1 × 4   ㉜ 7 × 1   ㉝ 1 × 1   ㉞ 9 × 1   ㉟ 1 × 2   ㊱ 0 × 1   ㊲ 1 × 8   ㊳ 5 × 1   ㊴ 1 × 3   ㊵ 6 × 1

㊶ 1 × 5   ㊷ 3 × 1   ㊸ 1 × 6   ㊹ 0 × 1   ㊺ 1 × 9   ㊻ 8 × 1   ㊼ 1 × 4   ㊽ 2 × 1   ㊾ 1 × 7   ㊿ 1 × 1

Study the 2 Facts

$\begin{array}{r} 2 \\ \times\ 0 \\ \hline 0 \end{array}$	$\begin{array}{r} 2 \\ \times\ 1 \\ \hline 2 \end{array}$	$\begin{array}{r} 2 \\ \times\ 2 \\ \hline 4 \end{array}$	$\begin{array}{r} 2 \\ \times\ 3 \\ \hline 6 \end{array}$	$\begin{array}{r} 2 \\ \times\ 4 \\ \hline 8 \end{array}$
$\begin{array}{r} 2 \\ \times\ 5 \\ \hline 10 \end{array}$	$\begin{array}{r} 2 \\ \times\ 6 \\ \hline 12 \end{array}$	$\begin{array}{r} 2 \\ \times\ 7 \\ \hline 14 \end{array}$	$\begin{array}{r} 2 \\ \times\ 8 \\ \hline 16 \end{array}$	$\begin{array}{r} 2 \\ \times\ 9 \\ \hline 18 \end{array}$
$\begin{array}{r} 0 \\ \times\ 2 \\ \hline 0 \end{array}$	$\begin{array}{r} 1 \\ \times\ 2 \\ \hline 2 \end{array}$	$\begin{array}{r} 2 \\ \times\ 2 \\ \hline 4 \end{array}$	$\begin{array}{r} 3 \\ \times\ 2 \\ \hline 6 \end{array}$	$\begin{array}{r} 4 \\ \times\ 2 \\ \hline 8 \end{array}$
$\begin{array}{r} 5 \\ \times\ 2 \\ \hline 10 \end{array}$	$\begin{array}{r} 6 \\ \times\ 2 \\ \hline 12 \end{array}$	$\begin{array}{r} 7 \\ \times\ 2 \\ \hline 14 \end{array}$	$\begin{array}{r} 8 \\ \times\ 2 \\ \hline 16 \end{array}$	$\begin{array}{r} 9 \\ \times\ 2 \\ \hline 18 \end{array}$

Practice with 2

① 2 × 9

② 3 × 2

③ 2 × 6

④ 4 × 2

⑤ 2 × 7

⑥ 2 × 2

⑦ 2 × 0

⑧ 8 × 2

⑨ 2 × 5

⑩ 1 × 2

⑪ 2 × 7

⑫ 0 × 2

⑬ 2 × 2

⑭ 8 × 2

⑮ 2 × 3

⑯ 5 × 2

⑰ 2 × 1

⑱ 9 × 2

⑲ 2 × 4

⑳ 6 × 2

㉑ 2 × 8

㉒ 4 × 2

㉓ 2 × 1

㉔ 0 × 2

㉕ 2 × 9

㉖ 6 × 2

㉗ 2 × 5

㉘ 7 × 2

㉙ 2 × 2

㉚ 3 × 2

㉛ 2 × 3

㉜ 9 × 2

㉝ 2 × 5

㉞ 1 × 2

㉟ 2 × 2

㊱ 7 × 2

㊲ 2 × 8

㊳ 4 × 2

㊴ 2 × 6

㊵ 0 × 2

㊶ 2 × 4

㊷ 6 × 2

㊸ 2 × 0

㊹ 7 × 2

㊺ 2 × 1

㊻ 8 × 2

㊼ 2 × 3

㊽ 2 × 2

㊾ 2 × 9

㊿ 5 × 2

Study the 3 Facts

3 × 0 = 0	3 × 1 = 3	3 × 2 = 6	3 × 3 = 9	3 × 4 = 12
3 × 5 = 15	3 × 6 = 18	3 × 7 = 21	3 × 8 = 24	3 × 9 = 27
0 × 3 = 0	1 × 3 = 3	2 × 3 = 6	3 × 3 = 9	4 × 3 = 12
5 × 3 = 15	6 × 3 = 18	7 × 3 = 21	8 × 3 = 24	9 × 3 = 27

Practice with 3

① 3 × 3 ② 5 × 3 ③ 3 × 9 ④ 0 × 3 ⑤ 3 × 8 ⑥ 6 × 3 ⑦ 3 × 1 ⑧ 4 × 3 ⑨ 3 × 2 ⑩ 7 × 3

⑪ 3 × 8 ⑫ 1 × 3 ⑬ 3 × 6 ⑭ 4 × 3 ⑮ 3 × 5 ⑯ 2 × 3 ⑰ 3 × 7 ⑱ 3 × 3 ⑲ 3 × 0 ⑳ 9 × 3

㉑ 3 × 4 ㉒ 0 × 3 ㉓ 3 × 7 ㉔ 1 × 3 ㉕ 3 × 3 ㉖ 9 × 3 ㉗ 3 × 2 ㉘ 8 × 3 ㉙ 3 × 6 ㉚ 5 × 3

㉛ 3 × 5 ㉜ 3 × 3 ㉝ 3 × 2 ㉞ 7 × 3 ㉟ 3 × 6 ㊱ 8 × 3 ㊲ 3 × 4 ㊳ 0 × 3 ㊴ 3 × 9 ㊵ 1 × 3

㊶ 3 × 0 ㊷ 9 × 3 ㊸ 3 × 1 ㊹ 8 × 3 ㊺ 3 × 7 ㊻ 4 × 3 ㊼ 3 × 5 ㊽ 6 × 3 ㊾ 3 × 3 ㊿ 2 × 3

Study the 4 Facts

4	4	4	4	4
× 0	× 1	× 2	× 3	× 4
0	4	8	12	16

4	4	4	4	4
× 5	× 6	× 7	× 8	× 9
20	24	28	32	36

0	1	2	3	4
× 4	× 4	× 4	× 4	× 4
0	4	8	12	16

5	6	7	8	9
× 4	× 4	× 4	× 4	× 4
20	24	28	32	36

Practice with 4

① 4 × 0   ② 8 × 4   ③ 4 × 6   ④ 7 × 4   ⑤ 4 × 4   ⑥ 1 × 4   ⑦ 4 × 5   ⑧ 9 × 4   ⑨ 4 × 2   ⑩ 3 × 4

⑪ 4 × 4   ⑫ 5 × 4   ⑬ 4 × 1   ⑭ 9 × 4   ⑮ 4 × 8   ⑯ 2 × 4   ⑰ 4 × 3   ⑱ 0 × 4   ⑲ 4 × 7   ⑳ 6 × 4

㉑ 4 × 9   ㉒ 7 × 4   ㉓ 4 × 3   ㉔ 5 × 4   ㉕ 4 × 0   ㉖ 6 × 4   ㉗ 4 × 2   ㉘ 4 × 4   ㉙ 4 × 1   ㉚ 8 × 4

㉛ 4 × 8   ㉜ 0 × 4   ㉝ 4 × 2   ㉞ 3 × 4   ㉟ 4 × 1   ㊱ 4 × 4   ㊲ 4 × 9   ㊳ 7 × 4   ㊴ 4 × 6   ㊵ 5 × 4

㊶ 4 × 7   ㊷ 6 × 4   ㊸ 4 × 5   ㊹ 4 × 4   ㊺ 4 × 3   ㊻ 9 × 4   ㊼ 4 × 8   ㊽ 1 × 4   ㊾ 4 × 0   ㊿ 2 × 4

Study the 5 Facts

$$
\begin{array}{r} 5 \\ \times\ 0 \\ \hline 0 \end{array}
\qquad
\begin{array}{r} 5 \\ \times\ 1 \\ \hline 5 \end{array}
\qquad
\begin{array}{r} 5 \\ \times\ 2 \\ \hline 10 \end{array}
\qquad
\begin{array}{r} 5 \\ \times\ 3 \\ \hline 15 \end{array}
\qquad
\begin{array}{r} 5 \\ \times\ 4 \\ \hline 20 \end{array}
$$

$$
\begin{array}{r} 5 \\ \times\ 5 \\ \hline 25 \end{array}
\qquad
\begin{array}{r} 5 \\ \times\ 6 \\ \hline 30 \end{array}
\qquad
\begin{array}{r} 5 \\ \times\ 7 \\ \hline 35 \end{array}
\qquad
\begin{array}{r} 5 \\ \times\ 8 \\ \hline 40 \end{array}
\qquad
\begin{array}{r} 5 \\ \times\ 9 \\ \hline 45 \end{array}
$$

$$
\begin{array}{r} 0 \\ \times\ 5 \\ \hline 0 \end{array}
\qquad
\begin{array}{r} 1 \\ \times\ 5 \\ \hline 5 \end{array}
\qquad
\begin{array}{r} 2 \\ \times\ 5 \\ \hline 10 \end{array}
\qquad
\begin{array}{r} 3 \\ \times\ 5 \\ \hline 15 \end{array}
\qquad
\begin{array}{r} 4 \\ \times\ 5 \\ \hline 20 \end{array}
$$

$$
\begin{array}{r} 5 \\ \times\ 5 \\ \hline 25 \end{array}
\qquad
\begin{array}{r} 6 \\ \times\ 5 \\ \hline 30 \end{array}
\qquad
\begin{array}{r} 7 \\ \times\ 5 \\ \hline 35 \end{array}
\qquad
\begin{array}{r} 8 \\ \times\ 5 \\ \hline 40 \end{array}
\qquad
\begin{array}{r} 9 \\ \times\ 5 \\ \hline 45 \end{array}
$$

Practice with 5

①   ②   ③   ④   ⑤   ⑥   ⑦   ⑧   ⑨   ⑩

①	②	③	④	⑤	⑥	⑦	⑧	⑨	⑩
5	2	5	9	5	4	5	5	5	1
× 8	× 5	× 6	× 5	× 7	× 5	× 3	× 5	× 0	× 5

⑪	⑫	⑬	⑭	⑮	⑯	⑰	⑱	⑲	⑳
5	3	5	5	5	0	5	8	5	6
× 7	× 5	× 4	× 5	× 2	× 5	× 1	× 5	× 9	× 5

㉑	㉒	㉓	㉔	㉕	㉖	㉗	㉘	㉙	㉚
5	9	5	3	5	6	5	7	5	2
× 5	× 5	× 1	× 5	× 8	× 5	× 0	× 5	× 4	× 5

㉛	㉜	㉝	㉞	㉟	㊱	㊲	㊳	㊴	㊵
5	8	5	1	5	7	5	9	5	3
× 2	× 5	× 0	× 5	× 4	× 5	× 5	× 5	× 6	× 5

㊶	㊷	㊸	㊹	㊺	㊻	㊼	㊽	㊾	㊿
5	6	5	7	5	5	5	4	5	0
× 9	× 5	× 3	× 5	× 1	× 5	× 2	× 5	× 8	× 5

Study the 6 Facts

6	6	6	6	6
× 0	× 1	× 2	× 3	× 4
0	6	12	18	24

6	6	6	6	6
× 5	× 6	× 7	× 8	× 9
30	36	42	48	54

0	1	2	3	4
× 6	× 6	× 6	× 6	× 6
0	6	12	18	24

5	6	7	8	9
× 6	× 6	× 6	× 6	× 6
30	36	42	48	54

Practice with 6

① 6 × 1   ② 7 × 6   ③ 6 × 6   ④ 9 × 6   ⑤ 6 × 4   ⑥ 8 × 6   ⑦ 6 × 5   ⑧ 3 × 6   ⑨ 6 × 2   ⑩ 0 × 6

⑪ 6 × 4   ⑫ 5 × 6   ⑬ 6 × 8   ⑭ 3 × 6   ⑮ 6 × 7   ⑯ 2 × 6   ⑰ 6 × 0   ⑱ 1 × 6   ⑲ 6 × 9   ⑳ 6 × 6

㉑ 6 × 3   ㉒ 9 × 6   ㉓ 6 × 0   ㉔ 5 × 6   ㉕ 6 × 1   ㉖ 6 × 6   ㉗ 6 × 2   ㉘ 4 × 6   ㉙ 6 × 8   ㉚ 7 × 6

㉛ 6 × 7   ㉜ 1 × 6   ㉝ 6 × 2   ㉞ 0 × 6   ㉟ 6 × 8   ㊱ 4 × 6   ㊲ 6 × 3   ㊳ 9 × 6   ㊴ 6 × 6   ㊵ 5 × 6

㊶ 6 × 9   ㊷ 6 × 6   ㊸ 6 × 5   ㊹ 4 × 6   ㊺ 6 × 0   ㊻ 3 × 6   ㊼ 6 × 7   ㊽ 8 × 6   ㊾ 6 × 1   ㊿ 2 × 6

Study the 7 Facts

7	7	7	7	7
× 0	× 1	× 2	× 3	× 4
0	7	14	21	28
7	7	7	7	7
× 5	× 6	× 7	× 8	× 9
35	42	49	56	63
0	1	2	3	4
× 7	× 7	× 7	× 7	× 7
0	7	14	21	28
5	6	7	8	9
× 7	× 7	× 7	× 7	× 7
35	42	49	56	63

Practice with 7

① 7 × 2   ② 5 × 7   ③ 7 × 3   ④ 0 × 7   ⑤ 7 × 4   ⑥ 6 × 7   ⑦ 7 × 7   ⑧ 8 × 7   ⑨ 7 × 9   ⑩ 1 × 7

⑪ 7 × 4   ⑫ 7 × 7   ⑬ 7 × 6   ⑭ 8 × 7   ⑮ 7 × 5   ⑯ 9 × 7   ⑰ 7 × 1   ⑱ 2 × 7   ⑲ 7 × 0   ⑳ 3 × 7

㉑ 7 × 8   ㉒ 0 × 7   ㉓ 7 × 1   ㉔ 7 × 7   ㉕ 7 × 2   ㉖ 3 × 7   ㉗ 7 × 9   ㉘ 4 × 7   ㉙ 7 × 6   ㉚ 5 × 7

㉛ 7 × 5   ㉜ 2 × 7   ㉝ 7 × 9   ㉞ 1 × 7   ㉟ 7 × 6   ㊱ 4 × 7   ㊲ 7 × 8   ㊳ 0 × 7   ㊴ 7 × 3   ㊵ 7 × 7

㊶ 7 × 0   ㊷ 3 × 7   ㊸ 7 × 7   ㊹ 4 × 7   ㊺ 7 × 1   ㊻ 8 × 7   ㊼ 7 × 5   ㊽ 6 × 7   ㊾ 7 × 2   ㊿ 9 × 7

Study the 8 Facts

8 × 0 ――― 0	8 × 1 ――― 8	8 × 2 ――― 16	8 × 3 ――― 24	8 × 4 ――― 32
8 × 5 ――― 40	8 × 6 ――― 48	8 × 7 ――― 56	8 × 8 ――― 64	8 × 9 ――― 72
0 × 8 ――― 0	1 × 8 ――― 8	2 × 8 ――― 16	3 × 8 ――― 24	4 × 8 ――― 32
5 × 8 ――― 40	6 × 8 ――― 48	7 × 8 ――― 56	8 × 8 ――― 64	9 × 8 ――― 72

Practice with 8

① 8 × 5

② 6 × 8

③ 8 × 2

④ 9 × 8

⑤ 8 × 3

⑥ 4 × 8

⑦ 8 × 1

⑧ 0 × 8

⑨ 8 × 7

⑩ 8 × 8

⑪ 8 × 3

⑫ 1 × 8

⑬ 8 × 4

⑭ 0 × 8

⑮ 8 × 6

⑯ 7 × 8

⑰ 8 × 8

⑱ 5 × 8

⑲ 8 × 9

⑳ 2 × 8

㉑ 8 × 0

㉒ 9 × 8

㉓ 8 × 8

㉔ 1 × 8

㉕ 8 × 5

㉖ 2 × 8

㉗ 8 × 7

㉘ 3 × 8

㉙ 8 × 4

㉚ 6 × 8

㉛ 8 × 6

㉜ 5 × 8

㉝ 8 × 7

㉞ 8 × 8

㉟ 8 × 4

㊱ 3 × 8

㊲ 8 × 0

㊳ 9 × 8

㊴ 8 × 2

㊵ 1 × 8

㊶ 8 × 9

㊷ 2 × 8

㊸ 8 × 1

㊹ 3 × 8

㊺ 8 × 8

㊻ 0 × 8

㊼ 8 × 6

㊽ 4 × 8

㊾ 8 × 5

㊿ 7 × 8

Study the 9 Facts

9 × 0 = 0	9 × 1 = 9	9 × 2 = 18	9 × 3 = 27	9 × 4 = 36
9 × 5 = 45	9 × 6 = 54	9 × 7 = 63	9 × 8 = 72	9 × 9 = 81
0 × 9 = 0	1 × 9 = 9	2 × 9 = 18	3 × 9 = 27	4 × 9 = 36
5 × 9 = 45	6 × 9 = 54	7 × 9 = 63	8 × 9 = 72	9 × 9 = 81

Practice with 9

① 9 × 6   ② 9 × 9   ③ 9 × 0   ④ 2 × 9   ⑤ 9 × 8   ⑥ 5 × 9   ⑦ 9 × 7   ⑧ 3 × 9   ⑨ 9 × 4   ⑩ 1 × 9

⑪ 9 × 8   ⑫ 7 × 9   ⑬ 9 × 5   ⑭ 3 × 9   ⑮ 9 × 9   ⑯ 4 × 9   ⑰ 9 × 1   ⑱ 6 × 9   ⑲ 9 × 2   ⑳ 0 × 9

㉑ 9 × 3   ㉒ 2 × 9   ㉓ 9 × 1   ㉔ 7 × 9   ㉕ 9 × 6   ㉖ 0 × 9   ㉗ 9 × 4   ㉘ 8 × 9   ㉙ 9 × 5   ㉚ 9 × 9

㉛ 9 × 9   ㉜ 6 × 9   ㉝ 9 × 4   ㉞ 1 × 9   ㉟ 9 × 5   ㊱ 8 × 9   ㊲ 9 × 3   ㊳ 2 × 9   ㊴ 9 × 0   ㊵ 7 × 9

㊶ 9 × 2   ㊷ 0 × 9   ㊸ 9 × 7   ㊹ 8 × 9   ㊺ 9 × 1   ㊻ 3 × 9   ㊼ 9 × 9   ㊽ 5 × 9   ㊾ 9 × 6   ㊿ 4 × 9

Smaller Facts (0-5)

① 2 × 5

② 5 × 2

③ 0 × 4

④ 3 × 1

⑤ 1 × 0

⑥ 4 × 0

⑦ 3 × 3

⑧ 2 × 2

⑨ 1 × 3

⑩ 0 × 1

⑪ 0 × 0

⑫ 3 × 3

⑬ 2 × 0

⑭ 1 × 1

⑮ 4 × 5

⑯ 5 × 4

⑰ 0 × 3

⑱ 4 × 1

⑲ 5 × 4

⑳ 2 × 2

㉑ 5 × 1

㉒ 4 × 4

㉓ 1 × 3

㉔ 5 × 5

㉕ 3 × 2

㉖ 2 × 4

㉗ 1 × 2

㉘ 3 × 0

㉙ 0 × 5

㉚ 4 × 3

㉛ 4 × 2

㉜ 1 × 5

㉝ 3 × 1

㉞ 0 × 0

㉟ 2 × 4

㊱ 0 × 3

㊲ 4 × 0

㊳ 5 × 3

㊴ 2 × 5

㊵ 3 × 4

㊶ 3 × 4

㊷ 0 × 0

㊸ 4 × 2

㊹ 2 × 3

㊺ 5 × 0

㊻ 1 × 5

㊼ 2 × 1

㊽ 0 × 2

㊾ 3 × 5

㊿ 1 × 4

Smaller Facts (0-5)

① 5 × 1   ② 2 × 0   ③ 1 × 2   ④ 0 × 4   ⑤ 3 × 5   ⑥ 2 × 0   ⑦ 4 × 2   ⑧ 5 × 3   ⑨ 1 × 5   ⑩ 3 × 1

⑪ 3 × 1   ⑫ 4 × 5   ⑬ 5 × 4   ⑭ 2 × 3   ⑮ 1 × 0   ⑯ 4 × 3   ⑰ 0 × 1   ⑱ 3 × 4   ⑲ 2 × 2   ⑳ 5 × 0

㉑ 2 × 4   ㉒ 0 × 2   ㉓ 3 × 3   ㉔ 4 × 0   ㉕ 5 × 4   ㉖ 3 × 1   ㉗ 2 × 5   ㉘ 1 × 1   ㉙ 0 × 0   ㉚ 4 × 3

㉛ 4 × 1   ㉜ 5 × 2   ㉝ 4 × 5   ㉞ 1 × 4   ㉟ 0 × 3   ㊱ 5 × 0   ㊲ 3 × 2   ㊳ 0 × 5   ㊴ 3 × 0   ㊵ 2 × 4

㊶ 0 × 0   ㊷ 3 × 1   ㊸ 0 × 2   ㊹ 5 × 5   ㊺ 2 × 1   ㊻ 4 × 4   ㊼ 1 × 0   ㊽ 2 × 3   ㊾ 5 × 2   ㊿ 1 × 3

Smaller Facts (0-5)

① 5
× 4

② 3
× 2

③ 4
× 0

④ 2
× 5

⑤ 1
× 1

⑥ 0
× 1

⑦ 2
× 3

⑧ 5
× 2

⑨ 1
× 3

⑩ 4
× 5

⑪ 4
× 1

⑫ 2
× 3

⑬ 5
× 1

⑭ 1
× 5

⑮ 0
× 4

⑯ 3
× 0

⑰ 4
× 3

⑱ 0
× 5

⑲ 3
× 0

⑳ 5
× 2

㉑ 3
× 5

㉒ 0
× 0

㉓ 1
× 3

㉔ 3
× 4

㉕ 2
× 2

㉖ 5
× 0

㉗ 1
× 2

㉘ 2
× 1

㉙ 4
× 4

㉚ 0
× 3

㉛ 0
× 2

㉜ 1
× 4

㉝ 2
× 5

㉞ 4
× 1

㉟ 5
× 0

㊱ 4
× 3

㊲ 0
× 1

㊳ 3
× 3

㊴ 5
× 4

㊵ 2
× 0

㊶ 2
× 0

㊷ 4
× 1

㊸ 0
× 2

㊹ 5
× 3

㊺ 3
× 1

㊻ 1
× 4

㊼ 5
× 5

㊽ 4
× 2

㊾ 2
× 4

㊿ 1
× 0

Smaller Facts (0-5)

①   ②   ③   ④   ⑤   ⑥   ⑦   ⑧   ⑨   ⑩
    3   5   1   4   2   5   0   3   1   2
  × 5   × 1   × 2   × 0   × 4   × 1   × 2   × 3   × 4   × 5

⑪   ⑫   ⑬   ⑭   ⑮   ⑯   ⑰   ⑱   ⑲   ⑳
    2   0   3   5   1   0   4   2   5   3
  × 5   × 4   × 0   × 3   × 1   × 3   × 5   × 0   × 2   × 1

㉑   ㉒   ㉓   ㉔   ㉕   ㉖   ㉗   ㉘   ㉙   ㉚
    5   4   2   0   3   2   5   1   4   0
  × 0   × 2   × 3   × 1   × 0   × 5   × 4   × 5   × 1   × 3

㉛   ㉜   ㉝   ㉞   ㉟   ㊱   ㊲   ㊳   ㊴   ㊵
    0   3   0   1   4   3   2   4   2   5
  × 5   × 2   × 4   × 0   × 3   × 1   × 2   × 4   × 1   × 0

㊶   ㊷   ㊸   ㊹   ㊺   ㊻   ㊼   ㊽   ㊾   ㊿
    4   2   4   3   5   0   1   5   3   1
  × 1   × 5   × 2   × 4   × 5   × 0   × 1   × 3   × 2   × 3

Smaller Facts (0-5)

① 1 × 1   ② 4 × 4   ③ 3 × 5   ④ 5 × 0   ⑤ 2 × 2   ⑥ 0 × 2   ⑦ 5 × 3   ⑧ 1 × 4   ⑨ 2 × 3   ⑩ 3 × 0

⑪ 3 × 2   ⑫ 5 × 3   ⑬ 1 × 2   ⑭ 2 × 0   ⑮ 0 × 1   ⑯ 4 × 5   ⑰ 3 × 3   ⑱ 0 × 0   ⑲ 4 × 5   ⑳ 1 × 4

㉑ 4 × 0   ㉒ 0 × 5   ㉓ 2 × 3   ㉔ 4 × 1   ㉕ 5 × 4   ㉖ 1 × 5   ㉗ 2 × 4   ㉘ 5 × 2   ㉙ 3 × 1   ㉚ 0 × 3

㉛ 0 × 4   ㉜ 2 × 1   ㉝ 5 × 0   ㉞ 3 × 2   ㉟ 1 × 5   ㊱ 3 × 3   ㊲ 0 × 2   ㊳ 4 × 3   ㊴ 1 × 1   ㊵ 5 × 5

㊶ 5 × 5   ㊷ 3 × 2   ㊸ 0 × 4   ㊹ 1 × 3   ㊺ 4 × 2   ㊻ 2 × 1   ㊼ 1 × 0   ㊽ 3 × 4   ㊾ 5 × 1   ㊿ 2 × 5

Page 34

Smaller Facts (0-5)

①	②	③	④	⑤	⑥	⑦	⑧	⑨	⑩
4	1	2	3	5	1	0	4	2	5
× 0	× 2	× 4	× 5	× 1	× 2	× 4	× 3	× 1	× 0

⑪	⑫	⑬	⑭	⑮	⑯	⑰	⑱	⑲	⑳
5	0	4	1	2	0	3	5	1	4
× 0	× 1	× 5	× 3	× 2	× 3	× 0	× 5	× 4	× 2

㉑	㉒	㉓	㉔	㉕	㉖	㉗	㉘	㉙	㉚
1	3	5	0	4	5	1	2	3	0
× 5	× 4	× 3	× 2	× 5	× 0	× 1	× 0	× 2	× 3

㉛	㉜	㉝	㉞	㉟	㊱	㊲	㊳	㊴	㊵
0	4	0	2	3	4	5	3	5	1
× 0	× 4	× 1	× 5	× 3	× 2	× 4	× 1	× 2	× 5

㊶	㊷	㊸	㊹	㊺	㊻	㊼	㊽	㊾	㊿
3	5	3	4	1	0	2	1	4	2
× 2	× 0	× 4	× 1	× 0	× 5	× 2	× 3	× 4	× 3

Smaller Facts (0-5)

① 3 × 5   ② 0 × 4   ③ 4 × 1   ④ 5 × 0   ⑤ 2 × 2   ⑥ 1 × 2   ⑦ 5 × 3   ⑧ 3 × 4   ⑨ 2 × 3   ⑩ 4 × 0

⑪ 4 × 2   ⑫ 5 × 3   ⑬ 3 × 2   ⑭ 2 × 0   ⑮ 1 × 5   ⑯ 0 × 1   ⑰ 4 × 3   ⑱ 1 × 0   ⑲ 0 × 1   ⑳ 3 × 4

㉑ 0 × 0   ㉒ 1 × 1   ㉓ 2 × 3   ㉔ 0 × 5   ㉕ 5 × 4   ㉖ 3 × 1   ㉗ 2 × 4   ㉘ 5 × 2   ㉙ 4 × 5   ㉚ 1 × 3

㉛ 1 × 4   ㉜ 2 × 5   ㉝ 5 × 0   ㉞ 4 × 2   ㉟ 3 × 1   ㊱ 4 × 3   ㊲ 1 × 2   ㊳ 0 × 3   ㊴ 3 × 5   ㊵ 5 × 1

㊶ 5 × 1   ㊷ 4 × 2   ㊸ 1 × 4   ㊹ 3 × 3   ㊺ 0 × 2   ㊻ 2 × 5   ㊼ 3 × 0   ㊽ 4 × 4   ㊾ 5 × 5   ㊿ 2 × 1

Smaller Facts (0-5)

①   ②   ③   ④   ⑤   ⑥   ⑦   ⑧   ⑨   ⑩
  0   3   2   4   5   3   1   0   2   5
× 0   × 2   × 4   × 1   × 5   × 2   × 4   × 3   × 5   × 0
____   ____   ____   ____   ____   ____   ____   ____   ____   ____

⑪   ⑫   ⑬   ⑭   ⑮   ⑯   ⑰   ⑱   ⑲   ⑳
  5   1   0   3   2   1   4   5   3   0
× 0   × 5   × 1   × 3   × 2   × 3   × 0   × 1   × 4   × 2
____   ____   ____   ____   ____   ____   ____   ____   ____   ____

㉑   ㉒   ㉓   ㉔   ㉕   ㉖   ㉗   ㉘   ㉙   ㉚
  3   4   5   1   0   5   3   2   4   1
× 1   × 4   × 3   × 2   × 1   × 0   × 5   × 0   × 2   × 3
____   ____   ____   ____   ____   ____   ____   ____   ____   ____

㉛   ㉜   ㉝   ㉞   ㉟   ㊱   ㊲   ㊳   ㊴   ㊵
  1   0   1   2   4   0   5   4   5   3
× 0   × 4   × 5   × 1   × 3   × 2   × 4   × 5   × 2   × 1
____   ____   ____   ____   ____   ____   ____   ____   ____   ____

㊶   ㊷   ㊸   ㊹   ㊺   ㊻   ㊼   ㊽   ㊾   ㊿
  4   5   4   0   3   1   2   3   0   2
× 2   × 0   × 4   × 5   × 0   × 1   × 2   × 3   × 4   × 3
____   ____   ____   ____   ____   ____   ____   ____   ____   ____

Smaller Facts (0-5)

①
$\begin{array}{r} 0 \\ \times\, 0 \\ \hline \end{array}$

②
$\begin{array}{r} 3 \\ \times\, 4 \\ \hline \end{array}$

③
$\begin{array}{r} 1 \\ \times\, 3 \\ \hline \end{array}$

④
$\begin{array}{r} 5 \\ \times\, 2 \\ \hline \end{array}$

⑤
$\begin{array}{r} 2 \\ \times\, 5 \\ \hline \end{array}$

⑥
$\begin{array}{r} 4 \\ \times\, 5 \\ \hline \end{array}$

⑦
$\begin{array}{r} 5 \\ \times\, 1 \\ \hline \end{array}$

⑧
$\begin{array}{r} 0 \\ \times\, 4 \\ \hline \end{array}$

⑨
$\begin{array}{r} 2 \\ \times\, 1 \\ \hline \end{array}$

⑩
$\begin{array}{r} 1 \\ \times\, 2 \\ \hline \end{array}$

⑪
$\begin{array}{r} 1 \\ \times\, 5 \\ \hline \end{array}$

⑫
$\begin{array}{r} 5 \\ \times\, 1 \\ \hline \end{array}$

⑬
$\begin{array}{r} 0 \\ \times\, 5 \\ \hline \end{array}$

⑭
$\begin{array}{r} 2 \\ \times\, 2 \\ \hline \end{array}$

⑮
$\begin{array}{r} 4 \\ \times\, 0 \\ \hline \end{array}$

⑯
$\begin{array}{r} 3 \\ \times\, 3 \\ \hline \end{array}$

⑰
$\begin{array}{r} 1 \\ \times\, 1 \\ \hline \end{array}$

⑱
$\begin{array}{r} 4 \\ \times\, 2 \\ \hline \end{array}$

⑲
$\begin{array}{r} 3 \\ \times\, 3 \\ \hline \end{array}$

⑳
$\begin{array}{r} 0 \\ \times\, 4 \\ \hline \end{array}$

㉑
$\begin{array}{r} 3 \\ \times\, 2 \\ \hline \end{array}$

㉒
$\begin{array}{r} 4 \\ \times\, 3 \\ \hline \end{array}$

㉓
$\begin{array}{r} 2 \\ \times\, 1 \\ \hline \end{array}$

㉔
$\begin{array}{r} 3 \\ \times\, 0 \\ \hline \end{array}$

㉕
$\begin{array}{r} 5 \\ \times\, 4 \\ \hline \end{array}$

㉖
$\begin{array}{r} 0 \\ \times\, 3 \\ \hline \end{array}$

㉗
$\begin{array}{r} 2 \\ \times\, 4 \\ \hline \end{array}$

㉘
$\begin{array}{r} 5 \\ \times\, 5 \\ \hline \end{array}$

㉙
$\begin{array}{r} 1 \\ \times\, 0 \\ \hline \end{array}$

㉚
$\begin{array}{r} 4 \\ \times\, 1 \\ \hline \end{array}$

㉛
$\begin{array}{r} 4 \\ \times\, 4 \\ \hline \end{array}$

㉜
$\begin{array}{r} 2 \\ \times\, 0 \\ \hline \end{array}$

㉝
$\begin{array}{r} 5 \\ \times\, 2 \\ \hline \end{array}$

㉞
$\begin{array}{r} 1 \\ \times\, 5 \\ \hline \end{array}$

㉟
$\begin{array}{r} 0 \\ \times\, 3 \\ \hline \end{array}$

㊱
$\begin{array}{r} 1 \\ \times\, 1 \\ \hline \end{array}$

㊲
$\begin{array}{r} 4 \\ \times\, 5 \\ \hline \end{array}$

㊳
$\begin{array}{r} 3 \\ \times\, 1 \\ \hline \end{array}$

㊴
$\begin{array}{r} 0 \\ \times\, 0 \\ \hline \end{array}$

㊵
$\begin{array}{r} 5 \\ \times\, 3 \\ \hline \end{array}$

㊶
$\begin{array}{r} 5 \\ \times\, 3 \\ \hline \end{array}$

㊷
$\begin{array}{r} 1 \\ \times\, 5 \\ \hline \end{array}$

㊸
$\begin{array}{r} 4 \\ \times\, 4 \\ \hline \end{array}$

㊹
$\begin{array}{r} 0 \\ \times\, 1 \\ \hline \end{array}$

㊺
$\begin{array}{r} 3 \\ \times\, 5 \\ \hline \end{array}$

㊻
$\begin{array}{r} 2 \\ \times\, 0 \\ \hline \end{array}$

㊼
$\begin{array}{r} 0 \\ \times\, 2 \\ \hline \end{array}$

㊽
$\begin{array}{r} 1 \\ \times\, 4 \\ \hline \end{array}$

㊾
$\begin{array}{r} 5 \\ \times\, 0 \\ \hline \end{array}$

㊿
$\begin{array}{r} 2 \\ \times\, 3 \\ \hline \end{array}$

Smaller Facts (0-5)

①	②	③	④	⑤	⑥	⑦	⑧	⑨	⑩
3	0	2	1	5	0	4	3	2	5
× 2	× 5	× 4	× 3	× 0	× 5	× 4	× 1	× 0	× 2

⑪	⑫	⑬	⑭	⑮	⑯	⑰	⑱	⑲	⑳
5	4	3	0	2	4	1	5	0	3
× 2	× 0	× 3	× 1	× 5	× 1	× 2	× 3	× 4	× 5

㉑	㉒	㉓	㉔	㉕	㉖	㉗	㉘	㉙	㉚
0	1	5	4	3	5	0	2	1	4
× 3	× 4	× 1	× 5	× 3	× 2	× 0	× 2	× 5	× 1

㉛	㉜	㉝	㉞	㉟	㊱	㊲	㊳	㊴	㊵
4	3	4	2	1	3	5	1	5	0
× 2	× 4	× 0	× 3	× 1	× 5	× 4	× 0	× 5	× 3

㊶	㊷	㊸	㊹	㊺	㊻	㊼	㊽	㊾	㊿
1	5	1	3	0	4	2	0	3	2
× 5	× 2	× 4	× 0	× 2	× 3	× 5	× 1	× 4	× 1

_____ _____ _____ _____ / 50
(name)   (date)   (time)   (score)

Smaller (0-5) Times Bigger (4-9)

①	②	③	④	⑤	⑥	⑦	⑧	⑨	⑩
1	9	3	6	4	9	1	6	1	5
× 8	× 0	× 4	× 5	× 5	× 0	× 7	× 4	× 8	× 2

⑪	⑫	⑬	⑭	⑮	⑯	⑰	⑱	⑲	⑳
9	5	1	6	2	5	2	4	5	5
× 3	× 4	× 7	× 0	× 7	× 1	× 9	× 0	× 8	× 4

㉑	㉒	㉓	㉔	㉕	㉖	㉗	㉘	㉙	㉚
2	7	0	8	0	6	5	8	2	9
× 4	× 3	× 5	× 4	× 7	× 3	× 4	× 3	× 6	× 1

㉛	㉜	㉝	㉞	㉟	㊱	㊲	㊳	㊴	㊵
3	6	1	8	4	4	2	5	4	7
× 9	× 5	× 4	× 2	× 5	× 4	× 8	× 5	× 7	× 0

㊶	㊷	㊸	㊹	㊺	㊻	㊼	㊽	㊾	㊿
4	9	5	7	5	6	0	5	3	4
× 6	× 2	× 8	× 3	× 9	× 2	× 7	× 1	× 5	× 5

Smaller (0-5) Times Bigger (4-9)

①	②	③	④	⑤	⑥	⑦	⑧	⑨	⑩
0	4	2	4	0	8	2	8	4	6
× 7	× 1	× 6	× 4	× 9	× 4	× 5	× 3	× 9	× 5

⑪	⑫	⑬	⑭	⑮	⑯	⑰	⑱	⑲	⑳
2	9	5	9	1	7	4	7	2	8
× 5	× 3	× 5	× 0	× 4	× 3	× 4	× 1	× 6	× 0

㉑	㉒	㉓	㉔	㉕	㉖	㉗	㉘	㉙	㉚
3	8	4	9	3	9	0	5	1	7
× 6	× 0	× 5	× 2	× 6	× 2	× 8	× 5	× 4	× 3

㉛	㉜	㉝	㉞	㉟	㊱	㊲	㊳	㊴	㊵
1	5	3	4	0	8	5	9	3	5
× 6	× 4	× 7	× 5	× 5	× 1	× 4	× 2	× 8	× 4

㊶	㊷	㊸	㊹	㊺	㊻	㊼	㊽	㊾	㊿
1	7	2	5	4	7	3	9	0	6
× 9	× 5	× 9	× 1	× 8	× 5	× 4	× 4	× 6	× 5

Smaller (0-5) Times Bigger (4-9)

①	②	③	④	⑤	⑥	⑦	⑧	⑨	⑩
4 × 8	5 × 5	3 × 7	6 × 0	2 × 9	5 × 5	4 × 4	6 × 2	4 × 8	9 × 1

⑪	⑫	⑬	⑭	⑮	⑯	⑰	⑱	⑲	⑳
0 × 7	9 × 2	4 × 4	6 × 5	1 × 4	9 × 4	1 × 5	7 × 5	0 × 8	5 × 3

㉑	㉒	㉓	㉔	㉕	㉖	㉗	㉘	㉙	㉚
1 × 7	4 × 3	5 × 9	8 × 2	5 × 4	6 × 3	0 × 7	8 × 3	1 × 6	5 × 4

㉛	㉜	㉝	㉞	㉟	㊱	㊲	㊳	㊴	㊵
3 × 5	6 × 0	4 × 7	8 × 1	2 × 9	7 × 2	1 × 8	9 × 0	2 × 4	4 × 5

㊶	㊷	㊸	㊹	㊺	㊻	㊼	㊽	㊾	㊿
2 × 6	5 × 1	0 × 8	4 × 3	0 × 5	6 × 1	5 × 4	9 × 4	3 × 9	7 × 0

Smaller (0-5) Times Bigger (4-9)

① 5 × 4
② 7 × 4
③ 1 × 6
④ 7 × 2
⑤ 5 × 5
⑥ 8 × 2
⑦ 1 × 9
⑧ 8 × 3
⑨ 2 × 5
⑩ 6 × 0

⑪ 1 × 9
⑫ 5 × 3
⑬ 0 × 9
⑭ 5 × 5
⑮ 4 × 7
⑯ 4 × 3
⑰ 2 × 7
⑱ 4 × 4
⑲ 1 × 6
⑳ 8 × 5

㉑ 3 × 6
㉒ 8 × 5
㉓ 2 × 9
㉔ 5 × 1
㉕ 3 × 6
㉖ 5 × 1
㉗ 5 × 8
㉘ 9 × 0
㉙ 4 × 7
㉚ 4 × 3

㉛ 0 × 7
㉜ 9 × 2
㉝ 3 × 4
㉞ 7 × 0
㉟ 5 × 9
㊱ 8 × 4
㊲ 4 × 6
㊳ 5 × 1
㊴ 3 × 8
㊵ 9 × 2

㊶ 4 × 5
㊷ 4 × 0
㊸ 1 × 5
㊹ 9 × 4
㊺ 2 × 8
㊻ 4 × 0
㊼ 3 × 7
㊽ 5 × 2
㊾ 5 × 6
㊿ 6 × 0

Smaller (0-5) Times Bigger (4-9)

①	②	③	④	⑤	⑥	⑦	⑧	⑨	⑩
2	4	3	7	2	4	4	7	4	8
× 7	× 5	× 9	× 1	× 8	× 5	× 6	× 2	× 5	× 0

⑪	⑫	⑬	⑭	⑮	⑯	⑰	⑱	⑲	⑳
1	8	4	7	0	8	0	9	1	4
× 9	× 2	× 6	× 5	× 6	× 4	× 4	× 5	× 5	× 3

㉑	㉒	㉓	㉔	㉕	㉖	㉗	㉘	㉙	㉚
0	6	5	5	5	7	1	5	0	4
× 9	× 3	× 8	× 2	× 6	× 3	× 9	× 3	× 7	× 4

㉛	㉜	㉝	㉞	㉟	㊱	㊲	㊳	㊴	㊵
3	7	4	5	2	9	0	8	2	6
× 4	× 1	× 9	× 0	× 8	× 2	× 5	× 1	× 6	× 5

㊶	㊷	㊸	㊹	㊺	㊻	㊼	㊽	㊾	㊿
4	4	1	6	1	7	5	8	3	9
× 5	× 0	× 5	× 3	× 4	× 0	× 6	× 4	× 8	× 1

Smaller (0-5) Times Bigger (4-9)

① 5 × 6	② 9 × 4	③ 0 × 7	④ 9 × 2	⑤ 5 × 4	⑥ 5 × 2	⑦ 0 × 8	⑧ 5 × 3	⑨ 2 × 4	⑩ 7 × 1

⑪ 0 × 8	⑫ 4 × 3	⑬ 1 × 8	⑭ 4 × 5	⑮ 4 × 9	⑯ 6 × 3	⑰ 2 × 9	⑱ 6 × 4	⑲ 0 × 7	⑳ 5 × 5

㉑ 3 × 7	㉒ 5 × 5	㉓ 2 × 8	㉔ 4 × 0	㉕ 3 × 7	㉖ 4 × 0	㉗ 5 × 5	㉘ 8 × 1	㉙ 4 × 9	㉚ 6 × 3

㉛ 1 × 9	㉜ 8 × 2	㉝ 3 × 6	㉞ 9 × 1	㉟ 5 × 8	㊱ 5 × 4	㊲ 4 × 7	㊳ 4 × 0	㊴ 3 × 5	㊵ 8 × 2

㊶ 4 × 4	㊷ 6 × 1	㊸ 0 × 4	㊹ 8 × 4	㊺ 2 × 5	㊻ 6 × 1	㊼ 3 × 9	㊽ 4 × 2	㊾ 5 × 7	㊿ 7 × 1

Smaller (0-5) Times Bigger (4-9)

①	②	③	④	⑤	⑥	⑦	⑧	⑨	⑩
0	9	1	6	2	9	0	6	0	5
× 7	× 5	× 8	× 3	× 5	× 5	× 4	× 2	× 7	× 4

⑪	⑫	⑬	⑭	⑮	⑯	⑰	⑱	⑲	⑳
3	5	0	6	4	5	4	8	3	9
× 8	× 2	× 4	× 5	× 4	× 0	× 9	× 5	× 7	× 1

㉑	㉒	㉓	㉔	㉕	㉖	㉗	㉘	㉙	㉚
4	4	5	7	5	6	3	7	4	9
× 8	× 1	× 5	× 2	× 4	× 1	× 8	× 1	× 6	× 0

㉛	㉜	㉝	㉞	㉟	㊱	㊲	㊳	㊴	㊵
1	6	0	7	2	8	4	5	2	4
× 9	× 3	× 8	× 4	× 5	× 2	× 7	× 3	× 4	× 5

㊶	㊷	㊸	㊹	㊺	㊻	㊼	㊽	㊾	㊿
2	9	3	4	3	6	5	5	1	8
× 6	× 4	× 7	× 1	× 9	× 4	× 4	× 0	× 5	× 3

Smaller (0-5) Times Bigger (4-9)

①	②	③	④	⑤	⑥	⑦	⑧	⑨	⑩
5	8	4	8	5	7	4	7	2	6
× 4	× 0	× 6	× 2	× 9	× 2	× 5	× 1	× 9	× 3

⑪	⑫	⑬	⑭	⑮	⑯	⑰	⑱	⑲	⑳
4	9	3	9	0	4	2	4	4	7
× 5	× 1	× 5	× 5	× 8	× 1	× 8	× 0	× 6	× 5

㉑	㉒	㉓	㉔	㉕	㉖	㉗	㉘	㉙	㉚
1	7	2	9	1	9	5	5	0	4
× 6	× 5	× 5	× 4	× 6	× 4	× 7	× 3	× 8	× 1

㉛	㉜	㉝	㉞	㉟	㊱	㊲	㊳	㊴	㊵
3	5	1	8	5	7	0	9	1	5
× 8	× 2	× 4	× 3	× 5	× 0	× 6	× 4	× 7	× 2

㊶	㊷	㊸	㊹	㊺	㊻	㊼	㊽	㊾	㊿
0	4	4	5	2	4	1	9	5	6
× 9	× 3	× 9	× 0	× 7	× 3	× 8	× 2	× 6	× 3

(name) (date) (time) (score)

Smaller (0-5) Times Bigger (4-9)

① 0 × 5 ② 7 × 3 ③ 4 × 6 ④ 8 × 2 ⑤ 1 × 9 ⑥ 7 × 3 ⑦ 0 × 4 ⑧ 8 × 1 ⑨ 0 × 5 ⑩ 9 × 5

⑪ 2 × 6 ⑫ 9 × 1 ⑬ 0 × 4 ⑭ 8 × 3 ⑮ 5 × 4 ⑯ 9 × 0 ⑰ 5 × 7 ⑱ 6 × 3 ⑲ 2 × 5 ⑳ 7 × 4

㉑ 5 × 6 ㉒ 4 × 4 ㉓ 3 × 9 ㉔ 5 × 1 ㉕ 3 × 4 ㉖ 8 × 4 ㉗ 2 × 6 ㉘ 5 × 4 ㉙ 5 × 8 ㉚ 7 × 0

㉛ 4 × 7 ㉜ 8 × 2 ㉝ 0 × 6 ㉞ 5 × 5 ㉟ 1 × 9 ㊱ 6 × 1 ㊲ 5 × 5 ㊳ 9 × 2 ㊴ 1 × 4 ㊵ 4 × 3

㊶ 1 × 8 ㊷ 7 × 5 ㊸ 2 × 5 ㊹ 4 × 4 ㊺ 2 × 7 ㊻ 8 × 5 ㊼ 3 × 4 ㊽ 9 × 0 ㊾ 4 × 9 ㊿ 6 × 2

Smaller (0-5) Times Bigger (4-9)

①   3   ②   6   ③   5   ④   6   ⑤   3   ⑥   5   ⑦   5   ⑧   5   ⑨   1   ⑩   8

× 4   × 0   × 8   × 1   × 7   × 1   × 9   × 4   × 7   × 2

⑪   5   ⑫   7   ⑬   2   ⑭   7   ⑮   0   ⑯   4   ⑰   1   ⑱   4   ⑲   5   ⑳   5

× 9   × 4   × 9   × 3   × 6   × 4   × 6   × 0   × 8   × 3

㉑   4   ㉒   5   ㉓   1   ㉔   7   ㉕   4   ㉖   7   ㉗   3   ㉘   9   ㉙   0   ㉚   4

× 8   × 3   × 9   × 5   × 8   × 5   × 5   × 2   × 6   × 4

㉛   2   ㉜   9   ㉝   4   ㉞   6   ㉟   3   ㊱   5   ㊲   0   ㊳   7   ㊴   4   ㊵   9

× 6   × 1   × 4   × 2   × 9   × 0   × 8   × 5   × 5   × 1

㊶   0   ㊷   4   ㊸   5   ㊹   9   ㊺   1   ㊻   4   ㊼   4   ㊽   7   ㊾   3   ㊿   8

× 7   × 2   × 7   × 0   × 5   × 2   × 6   × 1   × 8   × 2

Bigger Facts (4-9)

①	②	③	④	⑤	⑥	⑦	⑧	⑨	⑩
5	6	4	9	8	7	9	5	8	4
× 9	× 4	× 7	× 8	× 6	× 6	× 5	× 4	× 5	× 8

⑪	⑫	⑬	⑭	⑮	⑯	⑰	⑱	⑲	⑳
4	9	5	8	7	6	4	7	6	5
× 6	× 5	× 6	× 8	× 9	× 7	× 5	× 8	× 7	× 4

㉑	㉒	㉓	㉔	㉕	㉖	㉗	㉘	㉙	㉚
6	7	8	6	9	5	8	9	4	7
× 8	× 7	× 5	× 9	× 4	× 7	× 4	× 6	× 9	× 5

㉛	㉜	㉝	㉞	㉟	㊱	㊲	㊳	㊴	㊵
7	8	9	4	5	4	7	6	5	9
× 4	× 9	× 8	× 6	× 7	× 5	× 6	× 5	× 9	× 7

㊶	㊷	㊸	㊹	㊺	㊻	㊼	㊽	㊾	㊿
9	4	7	5	6	8	5	4	9	8
× 7	× 6	× 4	× 5	× 6	× 9	× 8	× 4	× 9	× 7

Bigger Facts (4-9)

_____ _____ _____ ____ / 50
(name)   (date)   (time) (score)

① 6 × 8   ② 5 × 6   ③ 8 × 4   ④ 4 × 7   ⑤ 9 × 9   ⑥ 5 × 6   ⑦ 7 × 4   ⑧ 6 × 5   ⑨ 8 × 9   ⑩ 9 × 8

⑪ 9 × 8   ⑫ 7 × 9   ⑬ 6 × 7   ⑭ 5 × 5   ⑮ 8 × 6   ⑯ 7 × 5   ⑰ 4 × 8   ⑱ 9 × 7   ⑲ 5 × 4   ⑳ 6 × 6

㉑ 5 × 7   ㉒ 4 × 4   ㉓ 9 × 5   ㉔ 7 × 6   ㉕ 6 × 7   ㉖ 9 × 8   ㉗ 5 × 9   ㉘ 8 × 8   ㉙ 4 × 6   ㉚ 7 × 5

㉛ 7 × 8   ㉜ 6 × 4   ㉝ 7 × 9   ㉞ 8 × 7   ㉟ 4 × 5   ㊱ 6 × 6   ㊲ 9 × 4   ㊳ 4 × 9   ㊴ 9 × 6   ㊵ 5 × 7

㊶ 4 × 6   ㊷ 9 × 8   ㊸ 4 × 4   ㊹ 6 × 9   ㊺ 5 × 8   ㊻ 7 × 7   ㊼ 8 × 6   ㊽ 5 × 5   ㊾ 6 × 4   ㊿ 8 × 5

Page 51

Bigger Facts (4-9)

① 8 ×5 ② 7 ×9 ③ 9 ×8 ④ 5 ×4 ⑤ 4 ×7 ⑥ 6 ×7 ⑦ 5 ×6 ⑧ 8 ×9 ⑨ 4 ×6 ⑩ 9 ×4

⑪ 9 ×7 ⑫ 5 ×6 ⑬ 8 ×7 ⑭ 4 ×4 ⑮ 6 ×5 ⑯ 7 ×8 ⑰ 9 ×6 ⑱ 6 ×4 ⑲ 7 ×8 ⑳ 8 ×9

㉑ 7 ×4 ㉒ 6 ×8 ㉓ 4 ×6 ㉔ 7 ×5 ㉕ 5 ×9 ㉖ 8 ×8 ㉗ 4 ×9 ㉘ 5 ×7 ㉙ 9 ×5 ㉚ 6 ×6

㉛ 6 ×9 ㉜ 4 ×5 ㉝ 5 ×4 ㉞ 9 ×7 ㉟ 8 ×8 ㊱ 9 ×6 ㊲ 6 ×7 ㊳ 7 ×6 ㊴ 8 ×5 ㊵ 5 ×8

㊶ 5 ×8 ㊷ 9 ×7 ㊸ 6 ×9 ㊹ 8 ×6 ㊺ 7 ×7 ㊻ 4 ×5 ㊼ 8 ×4 ㊽ 9 ×9 ㊾ 5 ×5 ㊿ 4 ×8

Bigger Facts (4-9)

① 7 × 4 ② 8 × 7 ③ 4 × 9 ④ 9 × 8 ⑤ 5 × 5 ⑥ 8 × 7 ⑦ 6 × 9 ⑧ 7 × 6 ⑨ 4 × 5 ⑩ 5 × 4

⑪ 5 × 4 ⑫ 6 × 5 ⑬ 7 × 8 ⑭ 8 × 6 ⑮ 4 × 7 ⑯ 6 × 6 ⑰ 9 × 4 ⑱ 5 × 8 ⑲ 8 × 9 ⑳ 7 × 7

㉑ 8 × 8 ㉒ 9 × 9 ㉓ 5 × 6 ㉔ 6 × 7 ㉕ 7 × 8 ㉖ 5 × 4 ㉗ 8 × 5 ㉘ 4 × 4 ㉙ 9 × 7 ㉚ 6 × 6

㉛ 6 × 4 ㉜ 7 × 9 ㉝ 6 × 5 ㉞ 4 × 8 ㉟ 9 × 6 ㊱ 7 × 7 ㊲ 5 × 9 ㊳ 9 × 5 ㊴ 5 × 7 ㊵ 8 × 8

㊶ 9 × 7 ㊷ 5 × 4 ㊸ 9 × 9 ㊹ 7 × 5 ㊺ 8 × 4 ㊻ 6 × 8 ㊼ 4 × 7 ㊽ 8 × 6 ㊾ 7 × 9 ㊿ 4 × 6

Bigger Facts (4-9)

① 7 × 4
② 9 × 8
③ 8 × 6
④ 5 × 7
⑤ 4 × 5
⑥ 6 × 5
⑦ 5 × 9
⑧ 7 × 8
⑨ 4 × 9
⑩ 8 × 7

⑪ 8 × 5
⑫ 5 × 9
⑬ 7 × 5
⑭ 4 × 7
⑮ 6 × 4
⑯ 9 × 6
⑰ 8 × 9
⑱ 6 × 7
⑲ 9 × 6
⑳ 7 × 8

㉑ 9 × 7
㉒ 6 × 6
㉓ 4 × 9
㉔ 9 × 4
㉕ 5 × 8
㉖ 7 × 6
㉗ 4 × 8
㉘ 5 × 5
㉙ 8 × 4
㉚ 6 × 9

㉛ 6 × 8
㉜ 4 × 4
㉝ 5 × 7
㉞ 8 × 5
㉟ 7 × 6
㊱ 8 × 9
㊲ 6 × 5
㊳ 9 × 9
㊴ 7 × 4
㊵ 5 × 6

㊶ 5 × 6
㊷ 8 × 5
㊸ 6 × 8
㊹ 7 × 9
㊺ 9 × 5
㊻ 4 × 4
㊼ 7 × 7
㊽ 8 × 8
㊾ 5 × 4
㊿ 4 × 6

Bigger Facts (4-9)

① 9 × 7   ② 7 × 5   ③ 4 × 8   ④ 8 × 6   ⑤ 5 × 4   ⑥ 7 × 5   ⑦ 6 × 8   ⑧ 9 × 9   ⑨ 4 × 4   ⑩ 5 × 7

⑪ 5 × 7   ⑫ 6 × 4   ⑬ 9 × 6   ⑭ 7 × 9   ⑮ 4 × 5   ⑯ 6 × 9   ⑰ 8 × 7   ⑱ 5 × 6   ⑲ 7 × 8   ⑳ 9 × 5

㉑ 7 × 6   ㉒ 8 × 8   ㉓ 5 × 9   ㉔ 6 × 5   ㉕ 9 × 6   ㉖ 5 × 7   ㉗ 7 × 4   ㉘ 4 × 7   ㉙ 8 × 5   ㉚ 6 × 9

㉛ 6 × 7   ㉜ 9 × 8   ㉝ 6 × 4   ㉞ 4 × 6   ㉟ 8 × 9   ㊱ 9 × 5   ㊲ 5 × 8   ㊳ 8 × 4   ㊴ 5 × 5   ㊵ 7 × 6

㊶ 8 × 5   ㊷ 5 × 7   ㊸ 8 × 8   ㊹ 9 × 4   ㊺ 7 × 7   ㊻ 6 × 6   ㊼ 4 × 5   ㊽ 7 × 9   ㊾ 9 × 8   ㊿ 4 × 9

Bigger Facts (4-9)

① 6 × 8 ② 5 × 5 ③ 9 × 9 ④ 4 × 7 ⑤ 8 × 4 ⑥ 7 × 4 ⑦ 4 × 6 ⑧ 6 × 5 ⑨ 8 × 6 ⑩ 9 × 7

⑪ 9 × 4 ⑫ 4 × 6 ⑬ 6 × 4 ⑭ 8 × 7 ⑮ 7 × 8 ⑯ 5 × 9 ⑰ 9 × 6 ⑱ 7 × 7 ⑲ 5 × 9 ⑳ 6 × 5

㉑ 5 × 7 ㉒ 7 × 9 ㉓ 8 × 6 ㉔ 5 × 8 ㉕ 4 × 5 ㉖ 6 × 9 ㉗ 8 × 5 ㉘ 4 × 4 ㉙ 9 × 8 ㉚ 7 × 6

㉛ 7 × 5 ㉜ 8 × 8 ㉝ 4 × 7 ㉞ 9 × 4 ㉟ 6 × 9 ㊱ 9 × 6 ㊲ 7 × 4 ㊳ 5 × 6 ㊴ 6 × 8 ㊵ 4 × 9

㊶ 4 × 9 ㊷ 9 × 4 ㊸ 7 × 5 ㊹ 6 × 6 ㊺ 5 × 4 ㊻ 8 × 8 ㊼ 6 × 7 ㊽ 9 × 5 ㊾ 4 × 8 ㊿ 8 × 9

Bigger Facts (4-9)

① 5 × 7   ② 6 × 4   ③ 8 × 5   ④ 9 × 9   ⑤ 4 × 8   ⑥ 6 × 4   ⑦ 7 × 5   ⑧ 5 × 6   ⑨ 8 × 8   ⑩ 4 × 7

⑪ 4 × 7   ⑫ 7 × 8   ⑬ 5 × 9   ⑭ 6 × 6   ⑮ 8 × 4   ⑯ 7 × 6   ⑰ 9 × 7   ⑱ 4 × 9   ⑲ 6 × 5   ⑳ 5 × 4

㉑ 6 × 9   ㉒ 9 × 5   ㉓ 4 × 6   ㉔ 7 × 4   ㉕ 5 × 9   ㉖ 4 × 7   ㉗ 6 × 8   ㉘ 8 × 7   ㉙ 9 × 4   ㉚ 7 × 6

㉛ 7 × 7   ㉜ 5 × 5   ㉝ 7 × 8   ㉞ 8 × 9   ㉟ 9 × 6   ㊱ 5 × 4   ㊲ 4 × 5   ㊳ 9 × 8   ㊴ 4 × 4   ㊵ 6 × 9

㊶ 9 × 4   ㊷ 4 × 7   ㊸ 9 × 5   ㊹ 5 × 8   ㊺ 6 × 7   ㊻ 7 × 9   ㊼ 8 × 4   ㊽ 6 × 6   ㊾ 5 × 5   ㊿ 8 × 6

Bigger Facts (4-9)

① 9 × 6   ② 5 × 9   ③ 8 × 5   ④ 4 × 7   ⑤ 7 × 4   ⑥ 6 × 4   ⑦ 4 × 8   ⑧ 9 × 9   ⑨ 7 × 8   ⑩ 8 × 7

⑪ 8 × 4   ⑫ 4 × 8   ⑬ 9 × 4   ⑭ 7 × 7   ⑮ 6 × 6   ⑯ 5 × 5   ⑰ 8 × 8   ⑱ 6 × 7   ⑲ 5 × 5   ⑳ 9 × 9

㉑ 5 × 7   ㉒ 6 × 5   ㉓ 7 × 8   ㉔ 5 × 6   ㉕ 4 × 9   ㉖ 9 × 5   ㉗ 7 × 9   ㉘ 4 × 4   ㉙ 8 × 6   ㉚ 6 × 8

㉛ 6 × 9   ㉜ 7 × 6   ㉝ 4 × 7   ㉞ 8 × 4   ㉟ 9 × 5   ㊱ 8 × 8   ㊲ 6 × 4   ㊳ 5 × 8   ㊴ 9 × 6   ㊵ 4 × 5

㊶ 4 × 5   ㊷ 8 × 4   ㊸ 6 × 9   ㊹ 9 × 8   ㊺ 5 × 4   ㊻ 7 × 6   ㊼ 9 × 7   ㊽ 8 × 9   ㊾ 4 × 6   ㊿ 7 × 5

Bigger Facts (4-9)

① 5 × 7   ② 9 × 4   ③ 7 × 9   ④ 8 × 5   ⑤ 4 × 6   ⑥ 9 × 4   ⑦ 6 × 9   ⑧ 5 × 8   ⑨ 7 × 6   ⑩ 4 × 7

⑪ 4 × 7   ⑫ 6 × 6   ⑬ 5 × 5   ⑭ 9 × 8   ⑮ 7 × 4   ⑯ 6 × 8   ⑰ 8 × 7   ⑱ 4 × 5   ⑲ 9 × 9   ⑳ 5 × 4

㉑ 9 × 5   ㉒ 8 × 9   ㉓ 4 × 8   ㉔ 6 × 4   ㉕ 5 × 5   ㉖ 4 × 7   ㉗ 9 × 6   ㉘ 7 × 7   ㉙ 8 × 4   ㉚ 6 × 8

㉛ 6 × 7   ㉜ 5 × 9   ㉝ 6 × 6   ㉞ 7 × 5   ㉟ 8 × 8   ㊱ 5 × 4   ㊲ 4 × 9   ㊳ 8 × 6   ㊴ 4 × 4   ㊵ 9 × 5

㊶ 8 × 4   ㊷ 4 × 7   ㊸ 8 × 9   ㊹ 5 × 6   ㊺ 9 × 7   ㊻ 6 × 5   ㊼ 7 × 4   ㊽ 9 × 8   ㊾ 5 × 9   ㊿ 7 × 8

Mixed Facts (0-9)

① 7 × 9 ② 4 × 4 ③ 0 × 2 ④ 1 × 0 ⑤ 8 × 3 ⑥ 6 × 7 ⑦ 2 × 4 ⑧ 9 × 6 ⑨ 5 × 8 ⑩ 3 × 5

⑪ 3 × 8 ⑫ 9 × 3 ⑬ 8 × 5 ⑭ 6 × 6 ⑮ 5 × 2 ⑯ 0 × 6 ⑰ 7 × 1 ⑱ 4 × 3 ⑲ 2 × 2 ⑳ 1 × 9

㉑ 6 × 0 ㉒ 5 × 7 ㉓ 2 × 7 ㉔ 4 × 1 ㉕ 7 × 8 ㉖ 1 × 5 ㉗ 3 × 3 ㉘ 8 × 4 ㉙ 0 × 5 ㉚ 9 × 0

㉛ 0 × 7 ㉜ 1 × 2 ㉝ 3 × 0 ㉞ 9 × 8 ㉟ 4 × 5 ㊱ 8 × 0 ㊲ 6 × 9 ㊳ 5 × 1 ㊴ 7 × 6 ㊵ 2 × 3

㊶ 9 × 2 ㊷ 8 × 1 ㊸ 7 × 5 ㊹ 5 × 3 ㊺ 0 × 0 ㊻ 4 × 8 ㊼ 1 × 7 ㊽ 2 × 8 ㊾ 3 × 6 ㊿ 6 × 2

Mixed Facts (0-9)

① 1 × 3 ② 2 × 6 ③ 6 × 1 ④ 8 × 9 ⑤ 3 × 4 ⑥ 5 × 9 ⑦ 9 × 5 ⑧ 7 × 7 ⑨ 4 × 2 ⑩ 0 × 1

⑪ 5 × 5 ⑫ 7 × 0 ⑬ 9 × 7 ⑭ 0 × 4 ⑮ 2 × 9 ⑯ 3 × 2 ⑰ 4 × 6 ⑱ 6 × 8 ⑲ 1 × 1 ⑳ 8 × 8

㉑ 2 × 1 ㉒ 6 × 5 ㉓ 4 × 9 ㉔ 3 × 1 ㉕ 1 × 6 ㉖ 9 × 4 ㉗ 5 × 0 ㉘ 0 × 3 ㉙ 8 × 7 ㉚ 7 × 4

㉛ 4 × 0 ㉜ 0 × 9 ㉝ 1 × 8 ㉞ 2 × 5 ㉟ 6 × 4 ㊱ 7 × 3 ㊲ 8 × 2 ㊳ 3 × 7 ㊴ 9 × 9 ㊵ 5 × 6

㊶ 8 × 6 ㊷ 3 × 9 ㊸ 5 × 4 ㊹ 7 × 2 ㊺ 9 × 1 ㊻ 2 × 0 ㊼ 0 × 8 ㊽ 1 × 4 ㊾ 6 × 3 ㊿ 4 × 7

Mixed Facts (0-9)

① 6 × 2	② 4 × 4	③ 5 × 8	④ 0 × 1	⑤ 9 × 6	⑥ 3 × 5	⑦ 2 × 4	⑧ 7 × 9	⑨ 8 × 7	⑩ 1 × 3
⑪ 1 × 7	⑫ 7 × 6	⑬ 9 × 3	⑭ 3 × 9	⑮ 8 × 8	⑯ 5 × 9	⑰ 6 × 0	⑱ 4 × 6	⑲ 2 × 8	⑳ 0 × 2
㉑ 3 × 1	㉒ 8 × 5	㉓ 2 × 5	㉔ 4 × 0	㉕ 6 × 7	㉖ 0 × 3	㉗ 1 × 6	㉘ 9 × 4	㉙ 5 × 3	㉚ 7 × 1
㉛ 5 × 5	㉜ 0 × 8	㉝ 1 × 1	㉞ 7 × 7	㉟ 4 × 3	㊱ 9 × 1	㊲ 3 × 2	㊳ 8 × 0	㊴ 6 × 9	㊵ 2 × 6
㊶ 7 × 8	㊷ 9 × 0	㊸ 6 × 3	㊹ 8 × 6	㊺ 5 × 1	㊻ 4 × 7	㊼ 0 × 5	㊽ 2 × 7	㊾ 1 × 9	㊿ 3 × 8

Mixed Facts (0-9)

① 　　② 　　③ 　　④ 　　⑤ 　　⑥ 　　⑦ 　　⑧ 　　⑨ 　　⑩

0	2	3	9	1	8	7	6	4	5
×6	×9	×0	×2	×4	×2	×3	×5	×8	×0

⑪ 　　⑫ 　　⑬ 　　⑭ 　　⑮ 　　⑯ 　　⑰ 　　⑱ 　　⑲ 　　⑳

8	6	7	5	2	1	4	3	0	9
×3	×1	×5	×4	×2	×8	×9	×7	×0	×7

㉑ 　　㉒ 　　㉓ 　　㉔ 　　㉕ 　　㉖ 　　㉗ 　　㉘ 　　㉙ 　　㉚

2	3	4	1	0	7	8	5	9	6
×0	×3	×2	×0	×9	×4	×1	×6	×5	×4

㉛ 　　㉜ 　　㉝ 　　㉞ 　　㉟ 　　㊱ 　　㊲ 　　㊳ 　　㊴ 　　㊵

4	5	0	2	3	6	9	1	7	8
×1	×2	×7	×3	×4	×6	×8	×5	×2	×9

㊶ 　　㊷ 　　㊸ 　　㊹ 　　㊺ 　　㊻ 　　㊼ 　　㊽ 　　㊾ 　　㊿

9	1	8	6	7	2	5	0	3	4
×9	×2	×4	×8	×0	×1	×7	×4	×6	×5

Mixed Facts (0-9)

① 8 × 0   ② 2 × 1   ③ 7 × 4   ④ 3 × 6   ⑤ 0 × 7   ⑥ 4 × 8   ⑦ 1 × 1   ⑧ 9 × 9   ⑨ 5 × 2   ⑩ 6 × 3

⑪ 6 × 2   ⑫ 9 × 7   ⑬ 0 × 3   ⑭ 4 × 9   ⑮ 5 × 4   ⑯ 7 × 9   ⑰ 8 × 5   ⑱ 2 × 7   ⑲ 1 × 4   ⑳ 3 × 0

㉑ 4 × 6   ㉒ 5 × 8   ㉓ 1 × 8   ㉔ 2 × 5   ㉕ 8 × 2   ㉖ 3 × 3   ㉗ 6 × 7   ㉘ 0 × 1   ㉙ 7 × 3   ㉚ 9 × 6

㉛ 7 × 8   ㉜ 3 × 4   ㉝ 6 × 6   ㉞ 9 × 2   ㉟ 2 × 3   ㊱ 0 × 6   ㊲ 4 × 0   ㊳ 5 × 5   ㊴ 8 × 9   ㊵ 1 × 7

㊶ 9 × 4   ㊷ 0 × 5   ㊸ 8 × 3   ㊹ 5 × 7   ㊺ 7 × 6   ㊻ 2 × 2   ㊼ 3 × 8   ㊽ 1 × 2   ㊾ 6 × 9   ㊿ 4 × 4

Mixed Facts (0-9)

① 　 ② 　 ③ 　 ④ 　 ⑤ 　 ⑥ 　 ⑦ 　 ⑧ 　 ⑨ 　 ⑩
$$\begin{array}{r} 3 \\ \times\, 7 \\ \hline \end{array} \quad \begin{array}{r} 1 \\ \times\, 9 \\ \hline \end{array} \quad \begin{array}{r} 4 \\ \times\, 5 \\ \hline \end{array} \quad \begin{array}{r} 0 \\ \times\, 0 \\ \hline \end{array} \quad \begin{array}{r} 6 \\ \times\, 1 \\ \hline \end{array} \quad \begin{array}{r} 5 \\ \times\, 0 \\ \hline \end{array} \quad \begin{array}{r} 9 \\ \times\, 3 \\ \hline \end{array} \quad \begin{array}{r} 8 \\ \times\, 8 \\ \hline \end{array} \quad \begin{array}{r} 2 \\ \times\, 4 \\ \hline \end{array} \quad \begin{array}{r} 7 \\ \times\, 5 \\ \hline \end{array}$$

⑪ 　 ⑫ 　 ⑬ 　 ⑭ 　 ⑮ 　 ⑯ 　 ⑰ 　 ⑱ 　 ⑲ 　 ⑳
$$\begin{array}{r} 5 \\ \times\, 3 \\ \hline \end{array} \quad \begin{array}{r} 8 \\ \times\, 6 \\ \hline \end{array} \quad \begin{array}{r} 9 \\ \times\, 8 \\ \hline \end{array} \quad \begin{array}{r} 7 \\ \times\, 1 \\ \hline \end{array} \quad \begin{array}{r} 1 \\ \times\, 0 \\ \hline \end{array} \quad \begin{array}{r} 6 \\ \times\, 4 \\ \hline \end{array} \quad \begin{array}{r} 2 \\ \times\, 9 \\ \hline \end{array} \quad \begin{array}{r} 4 \\ \times\, 2 \\ \hline \end{array} \quad \begin{array}{r} 3 \\ \times\, 5 \\ \hline \end{array} \quad \begin{array}{r} 0 \\ \times\, 2 \\ \hline \end{array}$$

㉑ 　 ㉒ 　 ㉓ 　 ㉔ 　 ㉕ 　 ㉖ 　 ㉗ 　 ㉘ 　 ㉙ 　 ㉚
$$\begin{array}{r} 1 \\ \times\, 5 \\ \hline \end{array} \quad \begin{array}{r} 4 \\ \times\, 3 \\ \hline \end{array} \quad \begin{array}{r} 2 \\ \times\, 0 \\ \hline \end{array} \quad \begin{array}{r} 6 \\ \times\, 5 \\ \hline \end{array} \quad \begin{array}{r} 3 \\ \times\, 9 \\ \hline \end{array} \quad \begin{array}{r} 9 \\ \times\, 1 \\ \hline \end{array} \quad \begin{array}{r} 5 \\ \times\, 6 \\ \hline \end{array} \quad \begin{array}{r} 7 \\ \times\, 7 \\ \hline \end{array} \quad \begin{array}{r} 0 \\ \times\, 8 \\ \hline \end{array} \quad \begin{array}{r} 8 \\ \times\, 1 \\ \hline \end{array}$$

㉛ 　 ㉜ 　 ㉝ 　 ㉞ 　 ㉟ 　 ㊱ 　 ㊲ 　 ㊳ 　 ㊴ 　 ㊵
$$\begin{array}{r} 2 \\ \times\, 6 \\ \hline \end{array} \quad \begin{array}{r} 7 \\ \times\, 0 \\ \hline \end{array} \quad \begin{array}{r} 3 \\ \times\, 2 \\ \hline \end{array} \quad \begin{array}{r} 1 \\ \times\, 3 \\ \hline \end{array} \quad \begin{array}{r} 4 \\ \times\, 1 \\ \hline \end{array} \quad \begin{array}{r} 8 \\ \times\, 7 \\ \hline \end{array} \quad \begin{array}{r} 0 \\ \times\, 4 \\ \hline \end{array} \quad \begin{array}{r} 6 \\ \times\, 8 \\ \hline \end{array} \quad \begin{array}{r} 9 \\ \times\, 0 \\ \hline \end{array} \quad \begin{array}{r} 5 \\ \times\, 9 \\ \hline \end{array}$$

㊶ 　 ㊷ 　 ㊸ 　 ㊹ 　 ㊺ 　 ㊻ 　 ㊼ 　 ㊽ 　 ㊾ 　 ㊿
$$\begin{array}{r} 0 \\ \times\, 9 \\ \hline \end{array} \quad \begin{array}{r} 6 \\ \times\, 0 \\ \hline \end{array} \quad \begin{array}{r} 5 \\ \times\, 1 \\ \hline \end{array} \quad \begin{array}{r} 8 \\ \times\, 4 \\ \hline \end{array} \quad \begin{array}{r} 9 \\ \times\, 5 \\ \hline \end{array} \quad \begin{array}{r} 1 \\ \times\, 6 \\ \hline \end{array} \quad \begin{array}{r} 7 \\ \times\, 2 \\ \hline \end{array} \quad \begin{array}{r} 3 \\ \times\, 1 \\ \hline \end{array} \quad \begin{array}{r} 4 \\ \times\, 7 \\ \hline \end{array} \quad \begin{array}{r} 2 \\ \times\, 8 \\ \hline \end{array}$$

Mixed Facts (0-9)

① 2   ② 1   ③ 7   ④ 6   ⑤ 4   ⑥ 8   ⑦ 0   ⑧ 3   ⑨ 5   ⑩ 9
× 0   × 3   × 6   × 2   × 9   × 1   × 3   × 8   × 5   × 4

⑪ 9   ⑫ 3   ⑬ 4   ⑭ 8   ⑮ 5   ⑯ 7   ⑰ 2   ⑱ 1   ⑲ 0   ⑳ 6
× 5   × 9   × 4   × 8   × 6   × 8   × 7   × 9   × 6   × 0

㉑ 8   ㉒ 5   ㉓ 0   ㉔ 1   ㉕ 2   ㉖ 6   ㉗ 9   ㉘ 4   ㉙ 7   ㉚ 3
× 2   × 1   × 1   × 7   × 5   × 4   × 9   × 3   × 4   × 2

㉛ 7   ㉜ 6   ㉝ 9   ㉞ 3   ㉟ 1   ㊱ 4   ㊲ 8   ㊳ 5   ㊴ 2   ㊵ 0
× 1   × 6   × 2   × 5   × 4   × 2   × 0   × 7   × 8   × 9

㊶ 3   ㊷ 4   ㊸ 2   ㊹ 5   ㊺ 7   ㊻ 1   ㊼ 6   ㊽ 0   ㊾ 9   ㊿ 8
× 6   × 7   × 4   × 9   × 2   × 5   × 1   × 5   × 8   × 6

Mixed Facts (0-9)

① 6 × 9   ② 0 × 8   ③ 8 × 7   ④ 4 × 0   ⑤ 9 × 3   ⑥ 5 × 0   ⑦ 3 × 4   ⑧ 2 × 1   ⑨ 1 × 6   ⑩ 7 × 7

⑪ 5 × 4   ⑫ 2 × 2   ⑬ 3 × 1   ⑭ 7 × 3   ⑮ 0 × 0   ⑯ 9 × 6   ⑰ 1 × 8   ⑱ 8 × 5   ⑲ 6 × 7   ⑳ 4 × 5

㉑ 0 × 7   ㉒ 8 × 4   ㉓ 1 × 0   ㉔ 9 × 7   ㉕ 6 × 8   ㉖ 3 × 3   ㉗ 5 × 2   ㉘ 7 × 9   ㉙ 4 × 1   ㉚ 2 × 3

㉛ 1 × 2   ㉜ 7 × 0   ㉝ 6 × 5   ㉞ 0 × 4   ㉟ 8 × 3   ㊱ 2 × 9   ㊲ 4 × 6   ㊳ 9 × 1   ㊴ 3 × 0   ㊵ 5 × 8

㊶ 4 × 8   ㊷ 9 × 0   ㊸ 5 × 3   ㊹ 2 × 6   ㊺ 3 × 7   ㊻ 0 × 2   ㊼ 7 × 5   ㊽ 6 × 3   ㊾ 8 × 9   ㊿ 1 × 1

Mixed Facts (0-9)

① 4 × 9   ② 8 × 7   ③ 5 × 1   ④ 1 × 5   ⑤ 3 × 0   ⑥ 7 × 8   ⑦ 2 × 7   ⑧ 0 × 6   ⑨ 6 × 4   ⑩ 9 × 3

⑪ 9 × 4   ⑫ 0 × 0   ⑬ 3 × 3   ⑭ 7 × 6   ⑮ 6 × 1   ⑯ 5 × 6   ⑰ 4 × 2   ⑱ 8 × 0   ⑲ 2 × 1   ⑳ 1 × 9

㉑ 7 × 5   ㉒ 6 × 8   ㉓ 2 × 8   ㉔ 8 × 2   ㉕ 4 × 4   ㉖ 1 × 3   ㉗ 9 × 0   ㉘ 3 × 7   ㉙ 5 × 3   ㉚ 0 × 5

㉛ 5 × 8   ㉜ 1 × 1   ㉝ 9 × 5   ㉞ 0 × 4   ㉟ 8 × 3   ㊱ 3 × 5   ㊲ 7 × 9   ㊳ 6 × 2   ㊴ 4 × 6   ㊵ 2 × 0

㊶ 0 × 1   ㊷ 3 × 2   ㊸ 4 × 3   ㊹ 6 × 0   ㊺ 5 × 5   ㊻ 8 × 4   ㊼ 1 × 8   ㊽ 2 × 4   ㊾ 9 × 6   ㊿ 7 × 1

Mixed Facts (0-9)

①	②	③	④	⑤	⑥	⑦	⑧	⑨	⑩
1	2	7	3	9	6	0	4	8	5
× 0	× 6	× 2	× 9	× 7	× 9	× 3	× 8	× 1	× 2

⑪	⑫	⑬	⑭	⑮	⑯	⑰	⑱	⑲	⑳
6	4	0	5	2	9	8	7	1	3
× 3	× 5	× 8	× 7	× 9	× 1	× 6	× 4	× 2	× 4

㉑	㉒	㉓	㉔	㉕	㉖	㉗	㉘	㉙	㉚
2	7	8	9	1	0	6	5	3	4
× 2	× 3	× 9	× 2	× 6	× 7	× 5	× 0	× 8	× 7

㉛	㉜	㉝	㉞	㉟	㊱	㊲	㊳	㊴	㊵
8	5	1	2	7	4	3	9	0	6
× 5	× 9	× 4	× 3	× 7	× 0	× 1	× 8	× 9	× 6

㊶	㊷	㊸	㊹	㊺	㊻	㊼	㊽	㊾	㊿
3	9	6	4	0	2	5	1	7	8
× 6	× 9	× 7	× 1	× 2	× 5	× 4	× 7	× 0	× 8

No Zero or One (2-9)

①	②	③	④	⑤	⑥	⑦	⑧	⑨	⑩
9 × 3	6 × 8	3 × 4	6 × 7	7 × 5	8 × 3	4 × 6	2 × 8	8 × 2	5 × 9

⑪	⑫	⑬	⑭	⑮	⑯	⑰	⑱	⑲	⑳
7 × 8	8 × 7	2 × 5	5 × 4	4 × 9	3 × 5	7 × 2	3 × 6	9 × 5	6 × 6

㉑	㉒	㉓	㉔	㉕	㉖	㉗	㉘	㉙	㉚
4 × 2	3 × 3	9 × 2	6 × 8	9 × 3	2 × 6	5 × 8	8 × 2	7 × 7	8 × 5

㉛	㉜	㉝	㉞	㉟	㊱	㊲	㊳	㊴	㊵
6 × 8	5 × 6	7 × 3	8 × 5	3 × 7	4 × 2	6 × 4	9 × 6	2 × 9	3 × 8

㊶	㊷	㊸	㊹	㊺	㊻	㊼	㊽	㊾	㊿
2 × 2	4 × 7	9 × 9	5 × 6	6 × 3	5 × 9	3 × 7	7 × 2	8 × 4	9 × 7

No Zero or One (2-9)

① 8 × 6 ② 7 × 9 ③ 4 × 8 ④ 3 × 2 ⑤ 2 × 7 ⑥ 7 × 2 ⑦ 9 × 5 ⑧ 6 × 5 ⑨ 5 × 3 ⑩ 2 × 4

⑪ 6 × 4 ⑫ 2 × 6 ⑬ 5 × 5 ⑭ 7 × 4 ⑮ 9 × 9 ⑯ 4 × 7 ⑰ 8 × 9 ⑱ 4 × 4 ⑲ 3 × 9 ⑳ 9 × 7

㉑ 4 × 5 ㉒ 9 × 8 ㉓ 8 × 4 ㉔ 2 × 3 ㉕ 3 × 5 ㉖ 6 × 9 ㉗ 4 × 4 ㉘ 5 × 7 ㉙ 4 × 8 ㉚ 7 × 6

㉛ 3 × 9 ㉜ 6 × 9 ㉝ 2 × 7 ㉞ 4 × 6 ㉟ 5 × 2 ㊱ 9 × 4 ㊲ 7 × 5 ㊳ 8 × 3 ㊴ 6 × 2 ㊵ 2 × 5

㊶ 5 × 6 ㊷ 7 × 4 ㊸ 6 × 3 ㊹ 9 × 6 ㊺ 8 × 8 ㊻ 3 × 6 ㊼ 2 × 3 ㊽ 3 × 8 ㊾ 5 × 2 ㊿ 4 × 3

No Zero or One (2-9)

① 7 × 7

② 5 × 6

③ 4 × 5

④ 5 × 4

⑤ 9 × 2

⑥ 3 × 7

⑦ 6 × 8

⑧ 8 × 6

⑨ 3 × 3

⑩ 2 × 9

⑪ 9 × 6

⑫ 3 × 4

⑬ 8 × 2

⑭ 2 × 5

⑮ 6 × 9

⑯ 4 × 2

⑰ 9 × 3

⑱ 4 × 8

⑲ 7 × 2

⑳ 5 × 8

㉑ 6 × 3

㉒ 4 × 7

㉓ 7 × 3

㉔ 5 × 6

㉕ 7 × 7

㉖ 8 × 8

㉗ 2 × 6

㉘ 3 × 3

㉙ 9 × 4

㉚ 3 × 2

㉛ 5 × 6

㉜ 2 × 8

㉝ 9 × 7

㉞ 3 × 2

㉟ 4 × 4

㊱ 6 × 3

㊲ 5 × 5

㊳ 7 × 8

㊴ 8 × 9

㊵ 4 × 6

㊶ 8 × 3

㊷ 6 × 4

㊸ 7 × 9

㊹ 2 × 8

㊺ 5 × 7

㊻ 2 × 9

㊼ 4 × 4

㊽ 9 × 3

㊾ 3 × 5

㊿ 7 × 4

No Zero or One (2-9)

①
3
× 8

②
9
× 9

③
6
× 6

④
4
× 3

⑤
8
× 4

⑥
9
× 3

⑦
7
× 2

⑧
5
× 2

⑨
2
× 7

⑩
8
× 5

⑪
5
× 5

⑫
8
× 8

⑬
2
× 2

⑭
9
× 5

⑮
7
× 9

⑯
6
× 4

⑰
3
× 9

⑱
6
× 5

⑲
4
× 9

⑳
7
× 4

㉑
6
× 2

㉒
7
× 6

㉓
3
× 5

㉔
8
× 7

㉕
4
× 2

㉖
5
× 9

㉗
6
× 5

㉘
2
× 4

㉙
6
× 6

㉚
9
× 8

㉛
4
× 9

㉜
5
× 9

㉝
8
× 4

㉞
6
× 8

㉟
2
× 3

㊱
7
× 5

㊲
9
× 2

㊳
3
× 7

㊴
5
× 3

㊵
8
× 2

㊶
2
× 8

㊷
9
× 5

㊸
5
× 7

㊹
7
× 8

㊺
3
× 6

㊻
4
× 8

㊼
8
× 7

㊽
4
× 6

㊾
2
× 3

㊿
6
× 7

No Zero or One (2-9)

①	②	③	④	⑤	⑥	⑦	⑧	⑨	⑩
3 × 6	5 × 4	4 × 3	5 × 8	2 × 7	7 × 6	9 × 2	6 × 4	7 × 5	8 × 9

⑪	⑫	⑬	⑭	⑮	⑯	⑰	⑱	⑲	⑳
2 × 4	7 × 8	6 × 7	8 × 3	9 × 9	4 × 7	2 × 5	4 × 2	3 × 7	5 × 2

㉑	㉒	㉓	㉔	㉕	㉖	㉗	㉘	㉙	㉚
9 × 5	4 × 6	3 × 5	5 × 4	3 × 6	6 × 2	8 × 4	7 × 5	2 × 8	7 × 7

㉛	㉜	㉝	㉞	㉟	㊱	㊲	㊳	㊴	㊵
5 × 4	8 × 2	2 × 6	7 × 7	4 × 8	9 × 5	5 × 3	3 × 2	6 × 9	4 × 4

㊶	㊷	㊸	㊹	㊺	㊻	㊼	㊽	㊾	㊿
6 × 5	9 × 8	3 × 9	8 × 2	5 × 6	8 × 9	4 × 8	2 × 5	7 × 3	3 × 8

No Zero or One (2-9)

① 7 × 2 ② 2 × 9 ③ 9 × 4 ④ 4 × 5 ⑤ 6 × 8 ⑥ 2 × 5 ⑦ 3 × 7 ⑧ 5 × 7 ⑨ 8 × 6 ⑩ 6 × 3

⑪ 5 × 3 ⑫ 6 × 2 ⑬ 8 × 7 ⑭ 2 × 3 ⑮ 3 × 9 ⑯ 9 × 8 ⑰ 7 × 9 ⑱ 9 × 3 ⑲ 4 × 9 ⑳ 3 × 8

㉑ 9 × 7 ㉒ 3 × 4 ㉓ 7 × 3 ㉔ 6 × 6 ㉕ 4 × 7 ㉖ 5 × 9 ㉗ 9 × 3 ㉘ 8 × 8 ㉙ 9 × 4 ㉚ 2 × 2

㉛ 4 × 9 ㉜ 5 × 9 ㉝ 6 × 8 ㉞ 9 × 2 ㉟ 8 × 5 ㊱ 3 × 3 ㊲ 2 × 7 ㊳ 7 × 6 ㊴ 5 × 5 ㊵ 6 × 7

㊶ 8 × 2 ㊷ 2 × 3 ㊸ 5 × 6 ㊹ 3 × 2 ㊺ 7 × 4 ㊻ 4 × 2 ㊼ 6 × 6 ㊽ 4 × 4 ㊾ 8 × 5 ㊿ 9 × 6

No Zero or One (2-9)

① 8 ×5   ② 7 ×4   ③ 4 ×9   ④ 7 ×3   ⑤ 2 ×8   ⑥ 5 ×5   ⑦ 6 ×2   ⑧ 3 ×4   ⑨ 5 ×6   ⑩ 9 ×7

⑪ 2 ×4   ⑫ 5 ×3   ⑬ 3 ×8   ⑭ 9 ×9   ⑮ 6 ×7   ⑯ 4 ×8   ⑰ 2 ×6   ⑱ 4 ×2   ⑲ 8 ×8   ⑳ 7 ×2

㉑ 6 ×6   ㉒ 4 ×5   ㉓ 8 ×6   ㉔ 7 ×4   ㉕ 8 ×5   ㉖ 3 ×2   ㉗ 9 ×4   ㉘ 5 ×6   ㉙ 2 ×3   ㉚ 5 ×8

㉛ 7 ×4   ㉜ 9 ×2   ㉝ 2 ×5   ㉞ 5 ×8   ㉟ 4 ×3   ㊱ 6 ×6   ㊲ 7 ×9   ㊳ 8 ×2   ㊴ 3 ×7   ㊵ 4 ×4

㊶ 3 ×6   ㊷ 6 ×3   ㊸ 8 ×7   ㊹ 9 ×2   ㊺ 7 ×5   ㊻ 9 ×7   ㊼ 4 ×3   ㊽ 2 ×6   ㊾ 5 ×9   ㊿ 8 ×3 (name)　　　　　　(date)　(time)　(score)

No Zero or One (2-9)

① 5 × 2　② 2 × 7　③ 6 × 4　④ 4 × 6　⑤ 3 × 3　⑥ 2 × 6　⑦ 8 × 8　⑧ 7 × 8　⑨ 9 × 5　⑩ 3 × 9

⑪ 7 × 9　⑫ 3 × 2　⑬ 9 × 8　⑭ 2 × 9　⑮ 8 × 7　⑯ 6 × 3　⑰ 5 × 7　⑱ 6 × 9　⑲ 4 × 7　⑳ 8 × 3

㉑ 6 × 8　㉒ 8 × 4　㉓ 5 × 9　㉔ 3 × 5　㉕ 4 × 8　㉖ 7 × 7　㉗ 6 × 9　㉘ 9 × 3　㉙ 6 × 4　㉚ 2 × 2

㉛ 4 × 7　㉜ 7 × 7　㉝ 3 × 3　㉞ 6 × 2　㉟ 9 × 6　㊱ 8 × 9　㊲ 2 × 8　㊳ 5 × 5　㊴ 7 × 6　㊵ 3 × 8

㊶ 9 × 2　㊷ 2 × 9　㊸ 7 × 5　㊹ 8 × 2　㊺ 5 × 4　㊻ 4 × 2　㊼ 3 × 5　㊽ 4 × 4　㊾ 9 × 6　㊿ 6 × 5

No Zero or One (2-9)

① $\begin{array}{r} 5 \\ \times\, 9 \\ \hline \end{array}$ ② $\begin{array}{r} 3 \\ \times\, 6 \\ \hline \end{array}$ ③ $\begin{array}{r} 8 \\ \times\, 8 \\ \hline \end{array}$ ④ $\begin{array}{r} 3 \\ \times\, 4 \\ \hline \end{array}$ ⑤ $\begin{array}{r} 6 \\ \times\, 7 \\ \hline \end{array}$ ⑥ $\begin{array}{r} 4 \\ \times\, 9 \\ \hline \end{array}$ ⑦ $\begin{array}{r} 9 \\ \times\, 2 \\ \hline \end{array}$ ⑧ $\begin{array}{r} 7 \\ \times\, 6 \\ \hline \end{array}$ ⑨ $\begin{array}{r} 4 \\ \times\, 5 \\ \hline \end{array}$ ⑩ $\begin{array}{r} 2 \\ \times\, 3 \\ \hline \end{array}$

⑪ $\begin{array}{r} 6 \\ \times\, 6 \\ \hline \end{array}$ ⑫ $\begin{array}{r} 4 \\ \times\, 4 \\ \hline \end{array}$ ⑬ $\begin{array}{r} 7 \\ \times\, 7 \\ \hline \end{array}$ ⑭ $\begin{array}{r} 2 \\ \times\, 8 \\ \hline \end{array}$ ⑮ $\begin{array}{r} 9 \\ \times\, 3 \\ \hline \end{array}$ ⑯ $\begin{array}{r} 8 \\ \times\, 7 \\ \hline \end{array}$ ⑰ $\begin{array}{r} 6 \\ \times\, 5 \\ \hline \end{array}$ ⑱ $\begin{array}{r} 8 \\ \times\, 2 \\ \hline \end{array}$ ⑲ $\begin{array}{r} 5 \\ \times\, 7 \\ \hline \end{array}$ ⑳ $\begin{array}{r} 3 \\ \times\, 2 \\ \hline \end{array}$

㉑ $\begin{array}{r} 9 \\ \times\, 5 \\ \hline \end{array}$ ㉒ $\begin{array}{r} 8 \\ \times\, 9 \\ \hline \end{array}$ ㉓ $\begin{array}{r} 5 \\ \times\, 5 \\ \hline \end{array}$ ㉔ $\begin{array}{r} 3 \\ \times\, 6 \\ \hline \end{array}$ ㉕ $\begin{array}{r} 5 \\ \times\, 9 \\ \hline \end{array}$ ㉖ $\begin{array}{r} 7 \\ \times\, 2 \\ \hline \end{array}$ ㉗ $\begin{array}{r} 2 \\ \times\, 6 \\ \hline \end{array}$ ㉘ $\begin{array}{r} 4 \\ \times\, 5 \\ \hline \end{array}$ ㉙ $\begin{array}{r} 6 \\ \times\, 4 \\ \hline \end{array}$ ㉚ $\begin{array}{r} 4 \\ \times\, 7 \\ \hline \end{array}$

㉛ $\begin{array}{r} 3 \\ \times\, 6 \\ \hline \end{array}$ ㉜ $\begin{array}{r} 2 \\ \times\, 2 \\ \hline \end{array}$ ㉝ $\begin{array}{r} 6 \\ \times\, 9 \\ \hline \end{array}$ ㉞ $\begin{array}{r} 4 \\ \times\, 7 \\ \hline \end{array}$ ㉟ $\begin{array}{r} 8 \\ \times\, 4 \\ \hline \end{array}$ ㊱ $\begin{array}{r} 9 \\ \times\, 5 \\ \hline \end{array}$ ㊲ $\begin{array}{r} 3 \\ \times\, 8 \\ \hline \end{array}$ ㊳ $\begin{array}{r} 5 \\ \times\, 2 \\ \hline \end{array}$ ㊴ $\begin{array}{r} 7 \\ \times\, 3 \\ \hline \end{array}$ ㊵ $\begin{array}{r} 8 \\ \times\, 6 \\ \hline \end{array}$

㊶ $\begin{array}{r} 7 \\ \times\, 5 \\ \hline \end{array}$ ㊷ $\begin{array}{r} 9 \\ \times\, 4 \\ \hline \end{array}$ ㊸ $\begin{array}{r} 5 \\ \times\, 3 \\ \hline \end{array}$ ㊹ $\begin{array}{r} 2 \\ \times\, 2 \\ \hline \end{array}$ ㊺ $\begin{array}{r} 3 \\ \times\, 9 \\ \hline \end{array}$ ㊻ $\begin{array}{r} 2 \\ \times\, 3 \\ \hline \end{array}$ ㊼ $\begin{array}{r} 8 \\ \times\, 4 \\ \hline \end{array}$ ㊽ $\begin{array}{r} 6 \\ \times\, 5 \\ \hline \end{array}$ ㊾ $\begin{array}{r} 4 \\ \times\, 8 \\ \hline \end{array}$ ㊿ $\begin{array}{r} 5 \\ \times\, 4 \\ \hline \end{array}$

No Zero or One (2-9)

① 4 × 2

② 6 × 3

③ 9 × 6

④ 8 × 5

⑤ 7 × 4

⑥ 6 × 5

⑦ 5 × 7

⑧ 3 × 7

⑨ 2 × 9

⑩ 7 × 8

⑪ 3 × 8

⑫ 7 × 2

⑬ 2 × 7

⑭ 6 × 8

⑮ 5 × 3

⑯ 9 × 4

⑰ 4 × 3

⑱ 9 × 8

⑲ 8 × 3

⑳ 5 × 4

㉑ 9 × 7

㉒ 5 × 6

㉓ 4 × 8

㉔ 7 × 9

㉕ 8 × 7

㉖ 3 × 3

㉗ 9 × 8

㉘ 2 × 4

㉙ 9 × 6

㉚ 6 × 2

㉛ 8 × 3

㉜ 3 × 3

㉝ 7 × 4

㉞ 9 × 2

㉟ 2 × 5

㊱ 5 × 8

㊲ 6 × 7

㊳ 4 × 9

㊴ 3 × 5

㊵ 7 × 7

㊶ 2 × 2

㊷ 6 × 8

㊸ 3 × 9

㊹ 5 × 2

㊺ 4 × 6

㊻ 8 × 2

㊼ 7 × 9

㊽ 8 × 6

㊾ 2 × 5

㊿ 9 × 9

Study the 10 Facts

10 × 0 ――― 0	10 × 1 ――― 10	10 × 2 ――― 20	10 × 3 ――― 30	10 × 4 ――― 40
10 × 5 ――― 50	10 × 6 ――― 60	10 × 7 ――― 70	10 × 8 ――― 80	10 × 9 ――― 90
	10 ×10 ――― 100	10 ×11 ――― 110	10 ×12 ――― 120	
0 ×10 ――― 0	1 ×10 ――― 10	2 ×10 ――― 20	3 ×10 ――― 30	4 ×10 ――― 40
5 ×10 ――― 50	6 ×10 ――― 60	7 ×10 ――― 70	8 ×10 ――― 80	9 ×10 ――― 90
	10 ×10 ――― 100	11 ×10 ――― 110	12 ×10 ――― 120	

Practice with 10

①
```
  10
×11
```

②
```
   7
×10
```

③
```
  10
× 5
```

④
```
   9
×10
```

⑤
```
  10
× 6
```

⑥
```
  10
×10
```

⑦
```
  10
×12
```

⑧
```
   1
×10
```

⑨
```
  10
× 4
```

⑩
```
   3
×10
```

⑪
```
  10
× 4
```

⑫
```
   1
×10
```

⑬
```
  10
× 9
```

⑭
```
   0
×10
```

⑮
```
  10
×11
```

⑯
```
  12
×10
```

⑰
```
  10
× 5
```

⑱
```
   7
×10
```

⑲
```
  10
× 8
```

⑳
```
   2
×10
```

㉑
```
  10
× 6
```

㉒
```
   9
×10
```

㉓
```
  10
×10
```

㉔
```
   8
×10
```

㉕
```
  10
× 3
```

㉖
```
   2
×10
```

㉗
```
  10
×11
```

㉘
```
   0
×10
```

㉙
```
  10
× 5
```

㉚
```
   4
×10
```

㉛
```
  10
× 8
```

㉜
```
   0
×10
```

㉝
```
  10
× 7
```

㉞
```
  12
×10
```

㉟
```
  10
×10
```

㊱
```
   6
×10
```

㊲
```
  10
× 4
```

㊳
```
   5
×10
```

㊴
```
  10
× 2
```

㊵
```
   1
×10
```

㊶
```
  10
× 7
```

㊷
```
   3
×10
```

㊸
```
  10
×12
```

㊹
```
   2
×10
```

㊺
```
  10
× 1
```

㊻
```
   9
×10
```

㊼
```
  10
× 8
```

㊽
```
  10
×10
```

㊾
```
  10
×11
```

㊿
```
   5
×10
```

Study the 11 Facts

11 × 0 = 0	11 × 1 = 11	11 × 2 = 22	11 × 3 = 33	11 × 4 = 44

$$11 \times 0 = 0 \qquad 11 \times 1 = 11 \qquad 11 \times 2 = 22 \qquad 11 \times 3 = 33 \qquad 11 \times 4 = 44$$

$$11 \times 5 = 55 \qquad 11 \times 6 = 66 \qquad 11 \times 7 = 77 \qquad 11 \times 8 = 88 \qquad 11 \times 9 = 99$$

$$11 \times 10 = 110 \qquad 11 \times 11 = 121 \qquad 11 \times 12 = 132$$

$$0 \times 11 = 0 \qquad 1 \times 11 = 11 \qquad 2 \times 11 = 22 \qquad 3 \times 11 = 33 \qquad 4 \times 11 = 44$$

$$5 \times 11 = 55 \qquad 6 \times 11 = 66 \qquad 7 \times 11 = 77 \qquad 8 \times 11 = 88 \qquad 9 \times 11 = 99$$

$$10 \times 11 = 110 \qquad 11 \times 11 = 121 \qquad 12 \times 11 = 132$$

Practice with 11

① 11   ② 11   ③ 11   ④ 9   ⑤ 11   ⑥ 6   ⑦ 11   ⑧ 8   ⑨ 11   ⑩ 12
× 7   ×11   × 4   ×11   × 1   ×11   × 5   ×11   × 3   ×11

⑪ 11   ⑫ 8   ⑬ 11   ⑭ 2   ⑮ 11   ⑯ 5   ⑰ 11   ⑱ 11   ⑲ 11   ⑳ 10
× 3   ×11   × 9   ×11   × 7   ×11   × 4   ×11   × 0   ×11

㉑ 11   ㉒ 9   ㉓ 11   ㉔ 0   ㉕ 11   ㉖ 10   ㉗ 11   ㉘ 2   ㉙ 11   ㉚ 3
× 1   ×11   × 6   ×11   ×12   ×11   × 7   ×11   × 4   ×11

㉛ 11   ㉜ 2   ㉝ 11   ㉞ 5   ㉟ 11   ㊱ 1   ㊲ 11   ㊳ 4   ㊴ 11   ㊵ 8
× 0   ×11   ×11   ×11   × 6   ×11   × 3   ×11   ×10   ×11

㊶ 11   ㊷ 12   ㊸ 11   ㊹ 10   ㊺ 11   ㊻ 9   ㊼ 11   ㊽ 6   ㊾ 11   ㊿ 4
×11   ×11   × 5   ×11   × 8   ×11   × 0   ×11   × 7   ×11

Study the 12 Facts

12 × 0 —— 0	12 × 1 —— 12	12 × 2 —— 24	12 × 3 —— 36	12 × 4 —— 48
12 × 5 —— 60	12 × 6 —— 72	12 × 7 —— 84	12 × 8 —— 96	12 × 9 —— 108
	12 ×10 —— 120	12 ×11 —— 132	12 ×12 —— 144	
0 ×12 —— 0	1 ×12 —— 12	2 ×12 —— 24	3 ×12 —— 36	4 ×12 —— 48
5 ×12 —— 60	6 ×12 —— 72	7 ×12 —— 84	8 ×12 —— 96	9 ×12 —— 108
	10 ×12 —— 120	11 ×12 —— 132	12 ×12 —— 144	

Practice with 12

① 12 × 8   ② 10 ×12   ③ 12 × 1   ④ 12 ×12   ⑤ 12 × 0   ⑥ 2 ×12   ⑦ 12 ×11   ⑧ 5 ×12   ⑨ 12 × 6   ⑩ 3 ×12

⑪ 12 × 6   ⑫ 5 ×12   ⑬ 12 ×12   ⑭ 4 ×12   ⑮ 12 × 8   ⑯ 11 ×12   ⑰ 12 × 1   ⑱ 10 ×12   ⑲ 12 × 7   ⑳ 9 ×12

㉑ 12 × 0   ㉒ 12 ×12   ㉓ 12 × 2   ㉔ 7 ×12   ㉕ 12 × 3   ㉖ 9 ×12   ㉗ 12 × 8   ㉘ 4 ×12   ㉙ 12 × 1   ㉚ 6 ×12

㉛ 12 × 7   ㉜ 4 ×12   ㉝ 12 ×10   ㉞ 11 ×12   ㉟ 12 × 2   ㊱ 0 ×12   ㊲ 12 × 6   ㊳ 1 ×12   ㊴ 12 × 9   ㊵ 5 ×12

㊶ 12 ×10   ㊷ 3 ×12   ㊸ 12 ×11   ㊹ 9 ×12   ㊺ 12 × 5   ㊻ 12 ×12   ㊼ 12 × 7   ㊽ 2 ×12   ㊾ 12 × 8   ㊿ 1 ×12

Smaller (0-9) Times Bigger (6-12)

①	②	③	④	⑤	⑥	⑦	⑧	⑨	⑩
4	7	5	11	2	7	4	10	4	12
× 9	× 0	× 8	× 3	× 9	× 0	× 6	× 2	× 8	× 1

⑪	⑫	⑬	⑭	⑮	⑯	⑰	⑱	⑲	⑳
3	8	4	12	1	10	1	11	3	9
× 6	× 2	× 7	× 0	×12	× 4	× 9	× 0	×11	× 5

㉑	㉒	㉓	㉔	㉕	㉖	㉗	㉘	㉙	㉚
1	10	0	9	0	9	3	12	1	6
×11	× 5	×10	× 2	× 6	× 5	× 8	× 5	× 7	× 4

㉛	㉜	㉝	㉞	㉟	㊱	㊲	㊳	㊴	㊵
5	12	4	8	2	11	1	12	2	9
× 8	× 3	×12	× 1	× 7	× 2	×10	× 3	× 6	× 0

㊶	㊷	㊸	㊹	㊺	㊻	㊼	㊽	㊾	㊿
2	10	3	6	3	8	0	8	5	7
×12	× 1	× 6	× 5	× 9	× 1	× 7	× 4	×11	× 3

Smaller (0-9) Times Bigger (6-12)

① 0 × 6   ② 11 × 4   ③ 1 × 6   ④ 11 × 2   ⑤ 0 × 8   ⑥ 12 × 2   ⑦ 1 ×10   ⑧ 7 × 5   ⑨ 2 × 9   ⑩ 10 × 3

⑪ 1 × 7   ⑫ 9 × 5   ⑬ 3 × 8   ⑭ 10 × 0   ⑮ 4 ×10   ⑯ 6 × 5   ⑰ 2 × 9   ⑱ 11 × 4   ⑲ 1 ×12   ⑳ 11 × 0

㉑ 5 ×10   ㉒ 6 × 0   ㉓ 2 ×11   ㉔ 8 × 1   ㉕ 5 ×12   ㉖ 7 × 1   ㉗ 0 × 7   ㉘ 9 × 3   ㉙ 4 × 6   ㉚ 6 × 5

㉛ 3 ×11   ㉜ 7 × 2   ㉝ 5 × 9   ㉞ 10 × 3   ㉟ 0 ×11   ㊱ 9 × 4   ㊲ 4 ×12   ㊳ 6 × 1   ㊴ 5 ×12   ㊵ 8 × 2

㊶ 4 ×10   ㊷ 12 × 3   ㊸ 1 × 9   ㊹ 7 × 4   ㊺ 2 × 6   ㊻ 7 × 3   ㊼ 5 ×11   ㊽ 8 × 2   ㊾ 0 ×10   ㊿ 8 × 3

Smaller (0-9) Times Bigger (6-12)

①   ②   ③   ④   ⑤   ⑥   ⑦   ⑧   ⑨   ⑩
    0   6   3   7   2   6   0   9   0   10
×12   × 5   × 8   × 1   ×12   × 5   ×11   × 2   × 8   × 4

⑪   ⑫   ⑬   ⑭   ⑮   ⑯   ⑰   ⑱   ⑲   ⑳
    1   8   0   10   4   9   4   7   1   12
×11   × 2   × 6   × 5   ×10   × 0   ×12   × 5   × 7   × 3

㉑   ㉒   ㉓   ㉔   ㉕   ㉖   ㉗   ㉘   ㉙   ㉚
    4   9   5   12   5   12   1   10   4   11
× 7   × 3   × 9   × 2   ×11   × 3   × 8   × 3   × 6   × 0

㉛   ㉜   ㉝   ㉞   ㉟   ㊱   ㊲   ㊳   ㊴   ㊵
    3   10   0   8   2   7   4   10   2   12
× 8   × 1   ×10   × 4   × 6   × 2   × 9   × 1   ×11   × 5

㊶   ㊷   ㊸   ㊹   ㊺   ㊻   ㊼   ㊽   ㊾   ㊿
    2   9   1   11   1   8   5   8   3   6
×10   × 4   ×11   × 3   ×12   × 4   × 6   × 0   × 7   × 1

Smaller (0-9) Times Bigger (6-12)

①   5   ②   7   ③   4   ④   7   ⑤   5   ⑥   10   ⑦   4   ⑧   6   ⑨   2   ⑩   9

×11   × 0   ×11   × 2   × 8   × 2   × 9   × 3   ×12   × 1

⑪   4   ⑫   12   ⑬   1   ⑭   9   ⑮   0   ⑯   11   ⑰   2   ⑱   7   ⑲   4   ⑳   7

× 6   × 3   × 8   × 5   × 9   × 3   ×12   × 0   ×10   × 5

㉑   3   ㉒   11   ㉓   2   ㉔   8   ㉕   3   ㉖   6   ㉗   5   ㉘   12   ㉙   0   ㉚   11

× 9   × 5   × 7   × 4   ×10   × 4   × 6   × 1   ×11   × 3

㉛   1   ㉜   6   ㉝   3   ㉞   9   ㉟   5   ㊱   12   ㊲   0   ㊳   11   ㊴   3   ㊵   8

× 7   × 2   ×12   × 1   × 7   × 0   ×10   × 4   ×10   × 2

㊶   0   ㊷   10   ㊸   4   ㊹   6   ㊺   2   ㊻   6   ㊼   3   ㊽   8   ㊾   5   ㊿   8

× 9   × 1   ×12   × 0   ×11   × 1   × 7   × 2   × 9   × 1

Smaller (0-9) Times Bigger (6-12)

① 4 × 7
② 8 × 2
③ 3 ×10
④ 9 × 0
⑤ 5 × 7
⑥ 8 × 2
⑦ 4 ×12
⑧ 11 × 5
⑨ 4 ×10
⑩ 6 × 1

⑪ 0 ×12
⑫ 10 × 5
⑬ 4 × 8
⑭ 6 × 2
⑮ 1 × 6
⑯ 11 × 4
⑰ 1 × 7
⑱ 9 × 2
⑲ 0 × 9
⑳ 7 × 3

㉑ 1 × 9
㉒ 11 × 3
㉓ 2 ×11
㉔ 7 × 5
㉕ 2 ×12
㉖ 7 × 3
㉗ 0 ×10
㉘ 6 × 3
㉙ 1 × 8
㉚ 12 × 4

㉛ 3 ×10
㉜ 6 × 0
㉝ 4 × 6
㉞ 10 × 1
㉟ 5 × 8
㊱ 9 × 5
㊲ 1 ×11
㊳ 6 × 0
㊴ 5 ×12
㊵ 7 × 2

㊶ 5 × 6
㊷ 11 × 1
㊸ 0 ×12
㊹ 12 × 3
㊺ 0 × 7
㊻ 10 × 1
㊼ 2 × 8
㊽ 10 × 4
㊾ 3 × 9
㊿ 8 × 0

Smaller (0-9) Times Bigger (6-12)

/ 50

(name) (date) (time) (score)

① 2 ×12
② 9 × 4
③ 1 ×12
④ 9 × 5
⑤ 2 ×10
⑥ 6 × 5
⑦ 1 ×11
⑧ 8 × 3
⑨ 5 × 7
⑩ 11 × 0

⑪ 1 × 8
⑫ 7 × 3
⑬ 0 ×10
⑭ 11 × 2
⑮ 4 ×11
⑯ 12 × 3
⑰ 5 × 7
⑱ 9 × 4
⑲ 1 × 6
⑳ 9 × 2

㉑ 3 ×11
㉒ 12 × 2
㉓ 5 × 9
㉔ 10 × 1
㉕ 3 × 6
㉖ 8 × 1
㉗ 2 × 8
㉘ 7 × 0
㉙ 4 ×12
㉚ 12 × 3

㉛ 0 × 9
㉜ 8 × 5
㉝ 3 × 7
㉞ 11 × 0
㉟ 2 × 9
㊱ 7 × 4
㊲ 4 × 6
㊳ 12 × 1
㊴ 3 × 6
㊵ 10 × 5

㊶ 4 ×11
㊷ 6 × 0
㊸ 1 × 7
㊹ 8 × 4
㊺ 5 ×12
㊻ 8 × 0
㊼ 3 × 9
㊽ 10 × 5
㊾ 2 ×11
㊿ 10 × 0

Page 91

Smaller (0-9) Times Bigger (6-12)

① 5 × 6   ② 10 × 2   ③ 0 × 7   ④ 8 × 4   ⑤ 3 × 6   ⑥ 10 × 2   ⑦ 5 × 9   ⑧ 12 × 3   ⑨ 5 × 7   ⑩ 11 × 1

⑪ 4 × 9   ⑫ 7 × 3   ⑬ 5 ×10   ⑭ 11 × 2   ⑮ 1 ×11   ⑯ 12 × 5   ⑰ 1 × 6   ⑱ 8 × 2   ⑲ 4 × 8   ⑳ 6 × 0

㉑ 1 × 8   ㉒ 12 × 0   ㉓ 2 ×12   ㉔ 6 × 3   ㉕ 2 × 9   ㉖ 6 × 0   ㉗ 4 × 7   ㉘ 11 × 0   ㉙ 1 ×10   ㉚ 9 × 5

㉛ 0 × 7   ㉜ 11 × 4   ㉝ 5 ×11   ㉞ 7 × 1   ㉟ 3 ×10   ㊱ 8 × 3   ㊲ 1 ×12   ㊳ 11 × 4   ㊴ 3 × 9   ㊵ 6 × 2

㊶ 3 ×11   ㊷ 12 × 1   ㊸ 4 × 9   ㊹ 9 × 0   ㊺ 4 × 6   ㊻ 7 × 1   ㊼ 2 ×10   ㊽ 7 × 5   ㊾ 0 × 8   ㊿ 10 × 4

Smaller (0-9) Times Bigger (6-12)

① 2 × 9   ② 8 × 5   ③ 1 × 9   ④ 8 × 3   ⑤ 2 × 7   ⑥ 11 × 3   ⑦ 1 ×12   ⑧ 10 × 0   ⑨ 3 × 6   ⑩ 12 × 4

⑪ 1 ×10   ⑫ 6 × 0   ⑬ 4 × 7   ⑭ 12 × 2   ⑮ 5 ×12   ⑯ 9 × 0   ⑰ 3 × 6   ⑱ 8 × 5   ⑲ 1 ×11   ⑳ 8 × 2

㉑ 0 ×12   ㉒ 9 × 2   ㉓ 3 × 8   ㉔ 7 × 1   ㉕ 0 ×11   ㉖ 10 × 1   ㉗ 2 ×10   ㉘ 6 × 4   ㉙ 5 × 9   ㉚ 9 × 0

㉛ 4 × 8   ㉜ 10 × 3   ㉝ 0 × 6   ㉞ 12 × 4   ㉟ 2 × 8   ㊱ 6 × 5   ㊲ 5 ×11   ㊳ 9 × 1   ㊴ 0 ×11   ㊵ 7 × 3

㊶ 5 ×12   ㊷ 11 × 4   ㊸ 1 × 6   ㊹ 10 × 5   ㊺ 3 × 9   ㊻ 10 × 4   ㊼ 0 × 8   ㊽ 7 × 3   ㊾ 2 ×12   ㊿ 7 × 4

Smaller (0-9) Times Bigger (6-12)

①	②	③	④	⑤	⑥	⑦	⑧	⑨	⑩
1	12	3	9	2	12	1	8	1	10
× 7	× 0	×11	× 5	× 7	× 0	× 6	× 2	×11	× 4

⑪	⑫	⑬	⑭	⑮	⑯	⑰	⑱	⑲	⑳
5	11	1	10	4	8	4	9	5	7
× 6	× 2	×12	× 0	×10	× 1	× 7	× 0	× 9	× 3

㉑	㉒	㉓	㉔	㉕	㉖	㉗	㉘	㉙	㉚
4	8	0	7	0	7	5	10	4	6
× 9	× 3	× 8	× 2	× 6	× 3	×11	× 3	×12	× 1

㉛	㉜	㉝	㉞	㉟	㊱	㊲	㊳	㊴	㊵
3	10	1	11	2	9	4	10	2	7
×11	× 5	×10	× 4	×12	× 2	× 8	× 5	× 6	× 0

㊶	㊷	㊸	㊹	㊺	㊻	㊼	㊽	㊾	㊿
2	8	5	6	5	11	0	11	3	12
×10	× 4	× 6	× 3	× 7	× 4	×12	× 1	× 9	× 5

Smaller (0-9) Times Bigger (6-12)

①	②	③	④	⑤	⑥	⑦	⑧	⑨	⑩
0	9	4	9	0	10	4	12	2	8
× 6	× 1	× 6	× 2	×11	× 2	× 8	× 3	× 7	× 5

⑪	⑫	⑬	⑭	⑮	⑯	⑰	⑱	⑲	⑳
4	7	5	8	1	6	2	9	4	9
×12	× 3	×11	× 0	× 8	× 3	× 7	× 1	×10	× 0

㉑	㉒	㉓	㉔	㉕	㉖	㉗	㉘	㉙	㉚
3	6	2	11	3	12	0	7	1	6
× 8	× 0	× 9	× 4	×10	× 4	×12	× 5	× 6	× 3

㉛	㉜	㉝	㉞	㉟	㊱	㊲	㊳	㊴	㊵
5	12	3	8	0	7	1	6	3	11
× 9	× 2	× 7	× 5	× 9	× 1	×10	× 4	×10	× 2

㊶	㊷	㊸	㊹	㊺	㊻	㊼	㊽	㊾	㊿
1	10	4	12	2	12	3	11	0	11
× 8	× 5	× 7	× 1	× 6	× 5	× 9	× 2	× 8	× 5

_____ _____ _____ _____ / 50
(name)   (date)   (time)   (score)

Bigger Facts (6-12)

① 9 × 7
② 10 × 9
③ 12 ×12
④ 7 × 8
⑤ 8 ×10
⑥ 6 ×11
⑦ 11 × 6
⑧ 8 × 7
⑨ 9 ×12
⑩ 7 ×10

⑪ 11 ×10
⑫ 8 ×11
⑬ 10 × 6
⑭ 11 ×12
⑮ 10 × 7
⑯ 12 × 9
⑰ 9 × 8
⑱ 12 × 6
⑲ 7 × 9
⑳ 6 ×11

㉑ 12 × 6
㉒ 7 ×12
㉓ 9 × 8
㉔ 6 × 9
㉕ 7 ×10
㉖ 8 × 6
㉗ 6 ×10
㉘ 10 × 8
㉙ 11 ×11
㉚ 12 × 7

㉛ 8 × 8
㉜ 7 × 9
㉝ 10 ×10
㉞ 9 ×11
㉟ 12 ×12
㊱ 11 × 8
㊲ 10 ×12
㊳ 6 × 6
㊴ 8 × 7
㊵ 9 × 9

㊶ 6 ×12
㊷ 11 × 7
㊸ 12 ×11
㊹ 8 × 6
㊺ 7 × 9
㊻ 9 ×10
㊼ 6 ×12
㊽ 7 × 7
㊾ 12 × 8
㊿ 10 ×11

Page 96

Bigger Facts (6-12)

① 8 ×10   ② 9 × 6   ③ 7 ×11   ④ 12 ×10   ⑤ 6 × 8   ⑥ 10 ×12   ⑦ 8 × 7   ⑧ 11 × 9   ⑨ 10 × 8   ⑩ 11 ×11

⑪ 7 × 9   ⑫ 12 × 8   ⑬ 8 ×12   ⑭ 10 × 9   ⑮ 11 ×10   ⑯ 7 × 7   ⑰ 12 ×11   ⑱ 9 × 7   ⑲ 6 × 6   ⑳ 9 × 6

㉑ 11 ×11   ㉒ 8 ×10   ㉓ 6 × 9   ㉔ 11 × 8   ㉕ 9 × 6   ㉖ 12 × 9   ㉗ 7 × 6   ㉘ 6 × 7   ㉙ 10 × 7   ㉚ 8 ×12

㉛ 10 ×12   ㉜ 12 × 8   ㉝ 7 ×10   ㉞ 6 × 7   ㉟ 12 ×10   ㊱ 11 × 8   ㊲ 8 ×12   ㊳ 11 ×11   ㊴ 9 × 9   ㊵ 7 × 6

㊶ 10 × 7   ㊷ 6 ×10   ㊸ 11 × 7   ㊹ 9 ×11   ㊺ 8 ×11   ㊻ 7 × 6   ㊼ 9 × 9   ㊽ 6 ×12   ㊾ 8 ×10   ㊿ 12 × 8

Bigger Facts (6-12)

① 12 × 6
② 7 ×11
③ 9 ×12
④ 6 ×10
⑤ 10 × 9
⑥ 8 × 7
⑦ 11 × 8
⑧ 10 × 6
⑨ 12 ×12
⑩ 6 × 9

⑪ 11 × 9
⑫ 10 × 7
⑬ 7 × 8
⑭ 11 ×12
⑮ 7 × 6
⑯ 9 ×11
⑰ 12 ×10
⑱ 9 × 8
⑲ 6 ×11
⑳ 8 × 7

㉑ 9 × 8
㉒ 6 ×12
㉓ 12 ×10
㉔ 8 ×11
㉕ 6 × 9
㉖ 10 × 8
㉗ 8 × 9
㉘ 7 ×10
㉙ 11 × 7
㉚ 9 × 6

㉛ 10 ×10
㉜ 6 ×11
㉝ 7 × 9
㉞ 12 × 7
㉟ 9 ×12
㊱ 11 ×10
㊲ 7 ×12
㊳ 8 × 8
㊴ 10 × 6
㊵ 12 ×11

㊶ 8 ×12
㊷ 11 × 6
㊸ 9 × 7
㊹ 10 × 8
㊺ 6 ×11
㊻ 12 × 9
㊼ 8 ×12
㊽ 6 × 6
㊾ 9 ×10
㊿ 7 × 7

Bigger Facts (6-12)

① 10 × 9
② 12 × 8
③ 6 × 7
④ 9 × 9
⑤ 8 ×10
⑥ 7 ×12
⑦ 10 × 6
⑧ 11 ×11
⑨ 7 ×10
⑩ 11 × 7

⑪ 6 ×11
⑫ 9 ×10
⑬ 10 ×12
⑭ 7 ×11
⑮ 11 × 9
⑯ 6 × 6
⑰ 9 × 7
⑱ 12 × 6
⑲ 8 × 8
⑳ 12 × 8

㉑ 11 × 7
㉒ 10 × 9
㉓ 8 ×11
㉔ 11 ×10
㉕ 12 × 8
㉖ 9 ×11
㉗ 6 × 8
㉘ 8 × 6
㉙ 7 × 6
㉚ 10 ×12

㉛ 10 ×12
㉜ 9 ×10
㉝ 6 × 9
㉞ 8 × 6
㉟ 9 × 9
㊱ 11 ×10
㊲ 7 ×12
㊳ 11 × 7
㊴ 12 ×11
㊵ 6 × 8

㊶ 7 × 6
㊷ 8 × 9
㊸ 11 × 6
㊹ 12 × 7
㊺ 10 × 7
㊻ 6 × 8
㊼ 12 ×11
㊽ 8 ×12
㊾ 10 × 9
㊿ 9 ×10

Bigger Facts (6-12)

① 8 ×11　② 12 × 9　③ 7 ×10　④ 10 × 7　⑤ 11 × 6　⑥ 6 ×12　⑦ 9 × 8　⑧ 11 ×11　⑨ 8 ×10　⑩ 10 × 6

⑪ 9 × 6　⑫ 11 ×12　⑬ 12 × 8　⑭ 9 ×10　⑮ 12 ×11　⑯ 7 × 9　⑰ 8 × 7　⑱ 7 × 8　⑲ 10 × 9　⑳ 6 ×12

㉑ 7 × 8　㉒ 10 ×10　㉓ 8 × 7　㉔ 6 × 9　㉕ 10 × 6　㉖ 11 × 8　㉗ 6 × 6　㉘ 12 × 7　㉙ 9 ×12　㉚ 7 ×11

㉛ 11 × 7　㉜ 10 × 9　㉝ 12 × 6　㉞ 8 ×12　㉟ 7 ×10　㊱ 9 × 7　㊲ 12 ×10　㊳ 6 × 8　㊴ 11 ×11　㊵ 8 × 9

㊶ 6 ×10　㊷ 9 ×11　㊸ 7 ×12　㊹ 11 × 8　㊺ 10 × 9　㊻ 8 × 6　㊼ 6 ×10　㊽ 10 ×11　㊾ 7 × 7　㊿ 12 ×12

Bigger Facts (6-12)

① 11 × 6
② 8 × 8
③ 10 ×12
④ 7 × 6
⑤ 6 × 7
⑥ 12 ×10
⑦ 11 ×11
⑧ 9 × 9
⑨ 12 × 7
⑩ 9 ×12

⑪ 10 × 9
⑫ 7 × 7
⑬ 11 ×10
⑭ 12 × 9
⑮ 9 × 6
⑯ 10 ×11
⑰ 7 ×12
⑱ 8 ×11
⑲ 6 × 8
⑳ 8 × 8

㉑ 9 ×12
㉒ 11 × 6
㉓ 6 × 9
㉔ 9 × 7
㉕ 8 × 8
㉖ 7 × 9
㉗ 10 × 8
㉘ 6 ×11
㉙ 12 ×11
㉚ 11 ×10

㉛ 11 ×10
㉜ 7 × 7
㉝ 10 × 6
㉞ 6 ×11
㉟ 7 × 6
㊱ 9 × 7
㊲ 12 ×10
㊳ 9 ×12
㊴ 8 × 9
㊵ 10 × 8

㊶ 12 ×11
㊷ 6 × 6
㊸ 9 ×11
㊹ 8 ×12
㊺ 11 ×12
㊻ 10 × 8
㊼ 8 × 9
㊽ 6 ×10
㊾ 11 × 6
㊿ 7 × 7

Bigger Facts (6-12)

① 6 ×12 ② 11 ×11 ③ 8 ×10 ④ 7 × 7 ⑤ 10 × 6 ⑥ 12 × 9 ⑦ 9 × 8 ⑧ 10 ×12 ⑨ 6 ×10 ⑩ 7 × 6

⑪ 9 × 6 ⑫ 10 × 9 ⑬ 11 × 8 ⑭ 9 ×10 ⑮ 11 ×12 ⑯ 8 ×11 ⑰ 6 × 7 ⑱ 8 × 8 ⑲ 7 ×11 ⑳ 12 × 9

㉑ 8 × 8 ㉒ 7 ×10 ㉓ 6 × 7 ㉔ 12 ×11 ㉕ 7 × 6 ㉖ 10 × 8 ㉗ 12 × 6 ㉘ 11 × 7 ㉙ 9 × 9 ㉚ 8 ×12

㉛ 10 × 7 ㉜ 7 ×11 ㉝ 11 × 6 ㉞ 6 × 9 ㉟ 8 ×10 ㊱ 9 × 7 ㊲ 11 ×10 ㊳ 12 × 8 ㊴ 10 ×12 ㊵ 6 ×11

㊶ 12 ×10 ㊷ 9 ×12 ㊸ 8 × 9 ㊹ 10 × 8 ㊺ 7 ×11 ㊻ 6 × 6 ㊼ 12 ×10 ㊽ 7 ×12 ㊾ 8 × 7 ㊿ 11 × 9

Bigger Facts (6-12)

① 10 × 6   ② 6 × 8   ③ 7 × 9   ④ 8 × 6   ⑤ 12 × 7   ⑥ 11 ×10   ⑦ 10 ×12   ⑧ 9 ×11   ⑨ 11 × 7   ⑩ 9 × 9

⑪ 7 ×11   ⑫ 8 × 7   ⑬ 10 ×10   ⑭ 11 ×11   ⑮ 9 × 6   ⑯ 7 ×12   ⑰ 8 × 9   ⑱ 6 ×12   ⑲ 12 × 8   ⑳ 6 × 8

㉑ 9 × 9   ㉒ 10 × 6   ㉓ 12 ×11   ㉔ 9 × 7   ㉕ 6 × 8   ㉖ 8 ×11   ㉗ 7 × 8   ㉘ 12 ×12   ㉙ 11 ×12   ㉚ 10 ×10

㉛ 10 ×10   ㉜ 8 × 7   ㉝ 7 × 6   ㉞ 12 ×12   ㉟ 8 × 6   ㊱ 9 × 7   ㊲ 11 ×10   ㊳ 9 × 9   ㊴ 6 ×11   ㊵ 7 × 8

㊶ 11 ×12   ㊷ 12 × 6   ㊸ 9 ×12   ㊹ 6 × 9   ㊺ 10 × 9   ㊻ 7 × 8   ㊼ 6 ×11   ㊽ 12 ×10   ㊾ 10 × 6   ㊿ 8 × 7

Bigger Facts (6-12)

①	②	③	④	⑤	⑥	⑦	⑧	⑨	⑩
11	10	7	9	12	6	8	12	11	9
×11	×10	×12	× 9	× 7	× 8	× 6	×11	×12	× 7

⑪	⑫	⑬	⑭	⑮	⑯	⑰	⑱	⑲	⑳
8	12	10	8	10	7	11	7	9	6
× 7	× 8	× 6	×12	×11	×10	× 9	× 6	×10	× 8

㉑	㉒	㉓	㉔	㉕	㉖	㉗	㉘	㉙	㉚
7	9	11	6	9	12	6	10	8	7
× 6	×12	× 9	×10	× 7	× 6	× 7	× 9	× 8	×11

㉛	㉜	㉝	㉞	㉟	㊱	㊲	㊳	㊴	㊵
12	9	10	11	7	8	10	6	12	11
× 9	×10	× 7	× 8	×12	× 9	×12	× 6	×11	×10

㊶	㊷	㊸	㊹	㊺	㊻	㊼	㊽	㊾	㊿
6	8	7	12	9	11	6	9	7	10
×12	×11	× 8	× 6	×10	× 7	×12	×11	× 9	× 8

Bigger Facts (6-12)

① 12 × 7   ② 11 × 6   ③ 9 × 8   ④ 7 × 7   ⑤ 6 × 9   ⑥ 10 ×12   ⑦ 12 ×11   ⑧ 8 ×10   ⑨ 10 × 9   ⑩ 8 × 8

⑪ 9 ×10   ⑫ 7 × 9   ⑬ 12 ×12   ⑭ 10 ×10   ⑮ 8 × 7   ⑯ 9 ×11   ⑰ 7 × 8   ⑱ 11 ×11   ⑲ 6 × 6   ⑳ 11 × 6

㉑ 8 × 8   ㉒ 12 × 7   ㉓ 6 ×10   ㉔ 8 × 9   ㉕ 11 × 6   ㉖ 7 ×10   ㉗ 9 × 6   ㉘ 6 ×11   ㉙ 10 ×11   ㉚ 12 ×12

㉛ 12 ×12   ㉜ 7 × 9   ㉝ 9 × 7   ㉞ 6 ×11   ㉟ 7 × 7   ㊱ 8 × 9   ㊲ 10 ×12   ㊳ 8 × 8   ㊴ 11 ×10   ㊵ 9 × 6

㊶ 10 ×11   ㊷ 6 × 7   ㊸ 8 ×11   ㊹ 11 × 8   ㊺ 12 × 8   ㊻ 9 × 6   ㊼ 11 ×10   ㊽ 6 ×12   ㊾ 12 × 7   ㊿ 7 × 9

Mixed Facts (0-12)

① $\begin{array}{r} 2 \\ \times\ 2 \\ \hline \end{array}$
② $\begin{array}{r} 9 \\ \times\ 9 \\ \hline \end{array}$
③ $\begin{array}{r} 12 \\ \times\ 5 \\ \hline \end{array}$
④ $\begin{array}{r} 7 \\ \times\ 3 \\ \hline \end{array}$
⑤ $\begin{array}{r} 4 \\ \times\ 6 \\ \hline \end{array}$
⑥ $\begin{array}{r} 11 \\ \times\ 0 \\ \hline \end{array}$
⑦ $\begin{array}{r} 3 \\ \times\ 1 \\ \hline \end{array}$
⑧ $\begin{array}{r} 0 \\ \times\ 8 \\ \hline \end{array}$
⑨ $\begin{array}{r} 10 \\ \times 11 \\ \hline \end{array}$
⑩ $\begin{array}{r} 6 \\ \times\ 4 \\ \hline \end{array}$

⑪ $\begin{array}{r} 0 \\ \times\ 7 \\ \hline \end{array}$
⑫ $\begin{array}{r} 3 \\ \times 10 \\ \hline \end{array}$
⑬ $\begin{array}{r} 11 \\ \times\ 1 \\ \hline \end{array}$
⑭ $\begin{array}{r} 8 \\ \times 12 \\ \hline \end{array}$
⑮ $\begin{array}{r} 9 \\ \times\ 2 \\ \hline \end{array}$
⑯ $\begin{array}{r} 6 \\ \times\ 0 \\ \hline \end{array}$
⑰ $\begin{array}{r} 7 \\ \times 11 \\ \hline \end{array}$
⑱ $\begin{array}{r} 1 \\ \times\ 3 \\ \hline \end{array}$
⑲ $\begin{array}{r} 5 \\ \times\ 6 \\ \hline \end{array}$
⑳ $\begin{array}{r} 12 \\ \times\ 9 \\ \hline \end{array}$

㉑ $\begin{array}{r} 5 \\ \times\ 9 \\ \hline \end{array}$
㉒ $\begin{array}{r} 4 \\ \times\ 3 \\ \hline \end{array}$
㉓ $\begin{array}{r} 1 \\ \times\ 8 \\ \hline \end{array}$
㉔ $\begin{array}{r} 10 \\ \times\ 4 \\ \hline \end{array}$
㉕ $\begin{array}{r} 12 \\ \times\ 7 \\ \hline \end{array}$
㉖ $\begin{array}{r} 9 \\ \times 12 \\ \hline \end{array}$
㉗ $\begin{array}{r} 0 \\ \times\ 5 \\ \hline \end{array}$
㉘ $\begin{array}{r} 6 \\ \times\ 2 \\ \hline \end{array}$
㉙ $\begin{array}{r} 8 \\ \times 10 \\ \hline \end{array}$
㉚ $\begin{array}{r} 2 \\ \times 11 \\ \hline \end{array}$

㉛ $\begin{array}{r} 8 \\ \times\ 8 \\ \hline \end{array}$
㉜ $\begin{array}{r} 10 \\ \times\ 5 \\ \hline \end{array}$
㉝ $\begin{array}{r} 7 \\ \times\ 6 \\ \hline \end{array}$
㉞ $\begin{array}{r} 2 \\ \times\ 0 \\ \hline \end{array}$
㉟ $\begin{array}{r} 3 \\ \times 11 \\ \hline \end{array}$
㊱ $\begin{array}{r} 1 \\ \times 10 \\ \hline \end{array}$
㊲ $\begin{array}{r} 5 \\ \times\ 4 \\ \hline \end{array}$
㊳ $\begin{array}{r} 4 \\ \times 12 \\ \hline \end{array}$
㊴ $\begin{array}{r} 0 \\ \times\ 1 \\ \hline \end{array}$
㊵ $\begin{array}{r} 11 \\ \times\ 7 \\ \hline \end{array}$

㊶ $\begin{array}{r} 1 \\ \times\ 4 \\ \hline \end{array}$
㊷ $\begin{array}{r} 11 \\ \times\ 2 \\ \hline \end{array}$
㊸ $\begin{array}{r} 6 \\ \times 12 \\ \hline \end{array}$
㊹ $\begin{array}{r} 4 \\ \times\ 1 \\ \hline \end{array}$
㊺ $\begin{array}{r} 8 \\ \times\ 9 \\ \hline \end{array}$
㊻ $\begin{array}{r} 7 \\ \times\ 0 \\ \hline \end{array}$
㊼ $\begin{array}{r} 12 \\ \times\ 6 \\ \hline \end{array}$
㊽ $\begin{array}{r} 9 \\ \times\ 7 \\ \hline \end{array}$
㊾ $\begin{array}{r} 2 \\ \times 10 \\ \hline \end{array}$
㊿ $\begin{array}{r} 3 \\ \times\ 3 \\ \hline \end{array}$

Mixed Facts (0-12)

① $\begin{array}{r} 6 \\ \times\ 1 \\ \hline \end{array}$ ② $\begin{array}{r} 2 \\ \times\ 6 \\ \hline \end{array}$ ③ $\begin{array}{r} 0 \\ \times 11 \\ \hline \end{array}$ ④ $\begin{array}{r} 9 \\ \times\ 0 \\ \hline \end{array}$ ⑤ $\begin{array}{r} 5 \\ \times\ 3 \\ \hline \end{array}$ ⑥ $\begin{array}{r} 12 \\ \times\ 2 \\ \hline \end{array}$ ⑦ $\begin{array}{r} 4 \\ \times\ 9 \\ \hline \end{array}$ ⑧ $\begin{array}{r} 10 \\ \times 12 \\ \hline \end{array}$ ⑨ $\begin{array}{r} 11 \\ \times\ 5 \\ \hline \end{array}$ ⑩ $\begin{array}{r} 7 \\ \times\ 8 \\ \hline \end{array}$

⑪ $\begin{array}{r} 7 \\ \times 12 \\ \hline \end{array}$ ⑫ $\begin{array}{r} 5 \\ \times\ 7 \\ \hline \end{array}$ ⑬ $\begin{array}{r} 3 \\ \times\ 9 \\ \hline \end{array}$ ⑭ $\begin{array}{r} 0 \\ \times 10 \\ \hline \end{array}$ ⑮ $\begin{array}{r} 10 \\ \times\ 8 \\ \hline \end{array}$ ⑯ $\begin{array}{r} 8 \\ \times\ 5 \\ \hline \end{array}$ ⑰ $\begin{array}{r} 9 \\ \times\ 0 \\ \hline \end{array}$ ⑱ $\begin{array}{r} 12 \\ \times\ 4 \\ \hline \end{array}$ ⑲ $\begin{array}{r} 6 \\ \times 11 \\ \hline \end{array}$ ⑳ $\begin{array}{r} 1 \\ \times\ 1 \\ \hline \end{array}$

㉑ $\begin{array}{r} 9 \\ \times\ 3 \\ \hline \end{array}$ ㉒ $\begin{array}{r} 8 \\ \times 11 \\ \hline \end{array}$ ㉓ $\begin{array}{r} 4 \\ \times 10 \\ \hline \end{array}$ ㉔ $\begin{array}{r} 11 \\ \times\ 8 \\ \hline \end{array}$ ㉕ $\begin{array}{r} 2 \\ \times\ 1 \\ \hline \end{array}$ ㉖ $\begin{array}{r} 0 \\ \times\ 4 \\ \hline \end{array}$ ㉗ $\begin{array}{r} 1 \\ \times\ 7 \\ \hline \end{array}$ ㉘ $\begin{array}{r} 5 \\ \times\ 5 \\ \hline \end{array}$ ㉙ $\begin{array}{r} 3 \\ \times\ 2 \\ \hline \end{array}$ ㉚ $\begin{array}{r} 10 \\ \times\ 6 \\ \hline \end{array}$

㉛ $\begin{array}{r} 11 \\ \times\ 6 \\ \hline \end{array}$ ㉜ $\begin{array}{r} 6 \\ \times\ 8 \\ \hline \end{array}$ ㉝ $\begin{array}{r} 10 \\ \times\ 2 \\ \hline \end{array}$ ㉞ $\begin{array}{r} 1 \\ \times\ 9 \\ \hline \end{array}$ ㉟ $\begin{array}{r} 7 \\ \times 10 \\ \hline \end{array}$ ㊱ $\begin{array}{r} 3 \\ \times\ 7 \\ \hline \end{array}$ ㊲ $\begin{array}{r} 12 \\ \times\ 3 \\ \hline \end{array}$ ㊳ $\begin{array}{r} 2 \\ \times 12 \\ \hline \end{array}$ ㊴ $\begin{array}{r} 4 \\ \times\ 4 \\ \hline \end{array}$ ㊵ $\begin{array}{r} 8 \\ \times\ 0 \\ \hline \end{array}$

㊶ $\begin{array}{r} 3 \\ \times\ 0 \\ \hline \end{array}$ ㊷ $\begin{array}{r} 12 \\ \times\ 1 \\ \hline \end{array}$ ㊸ $\begin{array}{r} 2 \\ \times\ 3 \\ \hline \end{array}$ ㊹ $\begin{array}{r} 5 \\ \times 12 \\ \hline \end{array}$ ㊺ $\begin{array}{r} 0 \\ \times\ 2 \\ \hline \end{array}$ ㊻ $\begin{array}{r} 4 \\ \times 11 \\ \hline \end{array}$ ㊼ $\begin{array}{r} 10 \\ \times 10 \\ \hline \end{array}$ ㊽ $\begin{array}{r} 11 \\ \times\ 9 \\ \hline \end{array}$ ㊾ $\begin{array}{r} 7 \\ \times\ 7 \\ \hline \end{array}$ ㊿ $\begin{array}{r} 9 \\ \times\ 5 \\ \hline \end{array}$

Mixed Facts (0-12)

① 9 × 8   ② 1 × 6   ③ 1 × 9   ④ 2 ×12   ⑤ 4 × 5   ⑥ 2 ×10   ⑦ 8 × 0   ⑧ 10 ×11   ⑨ 6 × 2   ⑩ 11 × 3

⑪ 10 × 4   ⑫ 8 × 7   ⑬ 2 × 0   ⑭ 5 × 1   ⑮ 1 × 8   ⑯ 11 ×10   ⑰ 2 × 2   ⑱ 12 ×12   ⑲ 0 × 5   ⑳ 1 × 6

㉑ 0 × 6   ㉒ 4 ×12   ㉓ 12 ×11   ㉔ 6 × 3   ㉕ 1 × 4   ㉖ 1 × 1   ㉗ 10 × 9   ㉘ 11 × 8   ㉙ 5 × 7   ㉚ 9 × 2

㉛ 5 ×11   ㉜ 6 × 9   ㉝ 2 × 5   ㉞ 9 ×10   ㉟ 8 × 2   ㊱ 12 × 7   ㊲ 0 × 3   ㊳ 4 ×12   ㊴ 10 × 0   ㊵ 2 × 4

㊶ 12 × 3   ㊷ 2 × 8   ㊸ 11 × 1   ㊹ 4 × 0   ㊺ 5 × 6   ㊻ 2 ×10   ㊼ 1 × 5   ㊽ 1 × 4   ㊾ 9 × 7   ㊿ 8 ×12

Mixed Facts (0-12)

① $\begin{array}{r} 11 \\ \times\ 0 \\ \hline \end{array}$
② $\begin{array}{r} 9 \\ \times\ 5 \\ \hline \end{array}$
③ $\begin{array}{r} 10 \\ \times\ 2 \\ \hline \end{array}$
④ $\begin{array}{r} 1 \\ \times 10 \\ \hline \end{array}$
⑤ $\begin{array}{r} 0 \\ \times 12 \\ \hline \end{array}$
⑥ $\begin{array}{r} 1 \\ \times\ 8 \\ \hline \end{array}$
⑦ $\begin{array}{r} 4 \\ \times\ 6 \\ \hline \end{array}$
⑧ $\begin{array}{r} 6 \\ \times\ 1 \\ \hline \end{array}$
⑨ $\begin{array}{r} 2 \\ \times\ 9 \\ \hline \end{array}$
⑩ $\begin{array}{r} 2 \\ \times 11 \\ \hline \end{array}$

⑪ $\begin{array}{r} 2 \\ \times\ 1 \\ \hline \end{array}$
⑫ $\begin{array}{r} 0 \\ \times\ 4 \\ \hline \end{array}$
⑬ $\begin{array}{r} 8 \\ \times\ 6 \\ \hline \end{array}$
⑭ $\begin{array}{r} 10 \\ \times\ 7 \\ \hline \end{array}$
⑮ $\begin{array}{r} 6 \\ \times 11 \\ \hline \end{array}$
⑯ $\begin{array}{r} 5 \\ \times\ 9 \\ \hline \end{array}$
⑰ $\begin{array}{r} 1 \\ \times 10 \\ \hline \end{array}$
⑱ $\begin{array}{r} 1 \\ \times\ 3 \\ \hline \end{array}$
⑲ $\begin{array}{r} 11 \\ \times\ 2 \\ \hline \end{array}$
⑳ $\begin{array}{r} 12 \\ \times\ 0 \\ \hline \end{array}$

㉑ $\begin{array}{r} 1 \\ \times 12 \\ \hline \end{array}$
㉒ $\begin{array}{r} 5 \\ \times\ 2 \\ \hline \end{array}$
㉓ $\begin{array}{r} 4 \\ \times\ 7 \\ \hline \end{array}$
㉔ $\begin{array}{r} 2 \\ \times 11 \\ \hline \end{array}$
㉕ $\begin{array}{r} 9 \\ \times\ 0 \\ \hline \end{array}$
㉖ $\begin{array}{r} 10 \\ \times\ 3 \\ \hline \end{array}$
㉗ $\begin{array}{r} 12 \\ \times\ 4 \\ \hline \end{array}$
㉘ $\begin{array}{r} 0 \\ \times\ 9 \\ \hline \end{array}$
㉙ $\begin{array}{r} 8 \\ \times\ 8 \\ \hline \end{array}$
㉚ $\begin{array}{r} 6 \\ \times\ 5 \\ \hline \end{array}$

㉛ $\begin{array}{r} 2 \\ \times\ 5 \\ \hline \end{array}$
㉜ $\begin{array}{r} 11 \\ \times 11 \\ \hline \end{array}$
㉝ $\begin{array}{r} 6 \\ \times\ 8 \\ \hline \end{array}$
㉞ $\begin{array}{r} 12 \\ \times\ 6 \\ \hline \end{array}$
㉟ $\begin{array}{r} 2 \\ \times\ 7 \\ \hline \end{array}$
㊱ $\begin{array}{r} 8 \\ \times\ 4 \\ \hline \end{array}$
㊲ $\begin{array}{r} 1 \\ \times 12 \\ \hline \end{array}$
㊳ $\begin{array}{r} 9 \\ \times\ 1 \\ \hline \end{array}$
㊴ $\begin{array}{r} 4 \\ \times\ 3 \\ \hline \end{array}$
㊵ $\begin{array}{r} 5 \\ \times 10 \\ \hline \end{array}$

㊶ $\begin{array}{r} 8 \\ \times 10 \\ \hline \end{array}$
㊷ $\begin{array}{r} 1 \\ \times\ 0 \\ \hline \end{array}$
㊸ $\begin{array}{r} 9 \\ \times 12 \\ \hline \end{array}$
㊹ $\begin{array}{r} 0 \\ \times\ 1 \\ \hline \end{array}$
㊺ $\begin{array}{r} 10 \\ \times\ 8 \\ \hline \end{array}$
㊻ $\begin{array}{r} 4 \\ \times\ 2 \\ \hline \end{array}$
㊼ $\begin{array}{r} 6 \\ \times\ 7 \\ \hline \end{array}$
㊽ $\begin{array}{r} 2 \\ \times\ 6 \\ \hline \end{array}$
㊾ $\begin{array}{r} 2 \\ \times\ 4 \\ \hline \end{array}$
㊿ $\begin{array}{r} 1 \\ \times\ 9 \\ \hline \end{array}$

Mixed Facts (0-12)

① 　4 　② 　0 　③ 　0 　④ 　3 　⑤ 11 　⑥ 　2 　⑦ 　5 　⑧ 　5 　⑨ 　8 　⑩ 10
　×10 　× 4 　×12 　× 3 　× 8 　× 5 　× 6 　× 9 　× 2 　× 7

⑪ 　5 　⑫ 　5 　⑬ 　2 　⑭ 　7 　⑮ 　0 　⑯ 10 　⑰ 　3 　⑱ 　9 　⑲ 12 　⑳ 　0
　× 1 　×11 　× 6 　× 0 　×10 　× 5 　× 2 　× 3 　× 8 　× 4

㉑ 12 　㉒ 11 　㉓ 　9 　㉔ 　8 　㉕ 　0 　㉖ 　0 　㉗ 　5 　㉘ 10 　㉙ 　7 　㉚ 　4
　× 4 　× 3 　× 9 　× 7 　× 1 　× 0 　×12 　×10 　×11 　× 2

㉛ 　7 　㉜ 　8 　㉝ 　3 　㉞ 　4 　㉟ 　5 　㊱ 　9 　㊲ 12 　㊳ 11 　㊴ 　5 　㊵ 　2
　× 9 　×12 　× 8 　× 5 　× 2 　×11 　× 7 　×12 　× 6 　× 1

㊶ 　9 　㊷ 　2 　㊸ 10 　㊹ 11 　㊺ 　7 　㊻ 　3 　㊼ 　0 　㊽ 　0 　㊾ 　4 　㊿ 　5
　× 7 　×10 　× 0 　× 6 　× 4 　× 5 　× 8 　× 1 　×11 　× 3

Mixed Facts (0-12)

① $\begin{array}{r} 10 \\ \times\ 6 \\ \hline \end{array}$
② $\begin{array}{r} 4 \\ \times\ 8 \\ \hline \end{array}$
③ $\begin{array}{r} 5 \\ \times\ 2 \\ \hline \end{array}$
④ $\begin{array}{r} 0 \\ \times\ 5 \\ \hline \end{array}$
⑤ $\begin{array}{r} 12 \\ \times\ 3 \\ \hline \end{array}$
⑥ $\begin{array}{r} 0 \\ \times 10 \\ \hline \end{array}$
⑦ $\begin{array}{r} 11 \\ \times\ 4 \\ \hline \end{array}$
⑧ $\begin{array}{r} 8 \\ \times\ 0 \\ \hline \end{array}$
⑨ $\begin{array}{r} 2 \\ \times 12 \\ \hline \end{array}$
⑩ $\begin{array}{r} 3 \\ \times\ 9 \\ \hline \end{array}$

⑪ $\begin{array}{r} 3 \\ \times\ 0 \\ \hline \end{array}$
⑫ $\begin{array}{r} 12 \\ \times\ 1 \\ \hline \end{array}$
⑬ $\begin{array}{r} 5 \\ \times\ 4 \\ \hline \end{array}$
⑭ $\begin{array}{r} 5 \\ \times 11 \\ \hline \end{array}$
⑮ $\begin{array}{r} 8 \\ \times\ 9 \\ \hline \end{array}$
⑯ $\begin{array}{r} 7 \\ \times 12 \\ \hline \end{array}$
⑰ $\begin{array}{r} 0 \\ \times\ 5 \\ \hline \end{array}$
⑱ $\begin{array}{r} 0 \\ \times\ 7 \\ \hline \end{array}$
⑲ $\begin{array}{r} 10 \\ \times\ 2 \\ \hline \end{array}$
⑳ $\begin{array}{r} 9 \\ \times\ 6 \\ \hline \end{array}$

㉑ $\begin{array}{r} 0 \\ \times\ 3 \\ \hline \end{array}$
㉒ $\begin{array}{r} 7 \\ \times\ 2 \\ \hline \end{array}$
㉓ $\begin{array}{r} 11 \\ \times 11 \\ \hline \end{array}$
㉔ $\begin{array}{r} 2 \\ \times\ 9 \\ \hline \end{array}$
㉕ $\begin{array}{r} 4 \\ \times\ 6 \\ \hline \end{array}$
㉖ $\begin{array}{r} 5 \\ \times\ 7 \\ \hline \end{array}$
㉗ $\begin{array}{r} 9 \\ \times\ 1 \\ \hline \end{array}$
㉘ $\begin{array}{r} 12 \\ \times 12 \\ \hline \end{array}$
㉙ $\begin{array}{r} 5 \\ \times 10 \\ \hline \end{array}$
㉚ $\begin{array}{r} 8 \\ \times\ 8 \\ \hline \end{array}$

㉛ $\begin{array}{r} 2 \\ \times\ 8 \\ \hline \end{array}$
㉜ $\begin{array}{r} 10 \\ \times\ 9 \\ \hline \end{array}$
㉝ $\begin{array}{r} 8 \\ \times 10 \\ \hline \end{array}$
㉞ $\begin{array}{r} 9 \\ \times\ 4 \\ \hline \end{array}$
㉟ $\begin{array}{r} 3 \\ \times 11 \\ \hline \end{array}$
㊱ $\begin{array}{r} 5 \\ \times\ 1 \\ \hline \end{array}$
㊲ $\begin{array}{r} 0 \\ \times\ 3 \\ \hline \end{array}$
㊳ $\begin{array}{r} 4 \\ \times\ 0 \\ \hline \end{array}$
㊴ $\begin{array}{r} 11 \\ \times\ 7 \\ \hline \end{array}$
㊵ $\begin{array}{r} 7 \\ \times\ 5 \\ \hline \end{array}$

㊶ $\begin{array}{r} 5 \\ \times\ 5 \\ \hline \end{array}$
㊷ $\begin{array}{r} 0 \\ \times\ 6 \\ \hline \end{array}$
㊸ $\begin{array}{r} 4 \\ \times\ 3 \\ \hline \end{array}$
㊹ $\begin{array}{r} 12 \\ \times\ 0 \\ \hline \end{array}$
㊺ $\begin{array}{r} 5 \\ \times 10 \\ \hline \end{array}$
㊻ $\begin{array}{r} 11 \\ \times\ 2 \\ \hline \end{array}$
㊼ $\begin{array}{r} 8 \\ \times 11 \\ \hline \end{array}$
㊽ $\begin{array}{r} 2 \\ \times\ 4 \\ \hline \end{array}$
㊾ $\begin{array}{r} 3 \\ \times\ 1 \\ \hline \end{array}$
㊿ $\begin{array}{r} 0 \\ \times 12 \\ \hline \end{array}$

Mixed Facts (0-12)

① 12 × 5   ② 11 ×12   ③ 1 × 8   ④ 9 × 2   ⑤ 10 ×11   ⑥ 6 × 0   ⑦ 1 × 7   ⑧ 0 × 9   ⑨ 0 × 6   ⑩ 5 × 4

⑪ 0 ×10   ⑫ 1 × 3   ⑬ 6 × 7   ⑭ 6 × 1   ⑮ 11 × 5   ⑯ 5 × 0   ⑰ 9 × 6   ⑱ 2 × 2   ⑲ 4 ×11   ⑳ 1 ×12

㉑ 4 ×12   ㉒ 10 × 2   ㉓ 2 × 9   ㉔ 0 × 4   ㉕ 1 ×10   ㉖ 11 × 1   ㉗ 0 × 8   ㉘ 5 × 5   ㉙ 6 × 3   ㉚ 12 × 6

㉛ 6 × 9   ㉜ 0 × 8   ㉝ 9 ×11   ㉞ 12 × 0   ㉟ 1 × 6   ㊱ 2 × 3   ㊲ 4 × 4   ㊳ 10 ×12   ㊴ 0 × 7   ㊵ 6 ×10

㊶ 2 × 4   ㊷ 6 × 5   ㊸ 5 × 1   ㊹ 10 × 7   ㊺ 6 ×12   ㊻ 9 × 0   ㊼ 1 ×11   ㊽ 11 ×10   ㊾ 12 × 3   ㊿ 1 × 2

Mixed Facts (0-12)

①	②	③	④	⑤	⑥	⑦	⑧	⑨	⑩
5	12	0	11	4	1	10	0	6	9
× 7	×11	× 6	× 0	× 2	× 5	×12	× 1	× 8	× 9

⑪	⑫	⑬	⑭	⑮	⑯	⑰	⑱	⑲	⑳
9	4	1	0	0	6	11	1	5	2
× 1	×10	×12	× 3	× 9	× 8	× 0	× 4	× 6	× 7

㉑	㉒	㉓	㉔	㉕	㉖	㉗	㉘	㉙	㉚
11	6	10	6	12	0	2	4	1	0
× 2	× 6	× 3	× 9	× 7	× 4	×10	× 8	× 5	×11

㉛	㉜	㉝	㉞	㉟	㊱	㊲	㊳	㊴	㊵
6	5	0	2	9	1	1	12	10	6
×11	× 9	× 5	×12	× 3	×10	× 2	× 1	× 4	× 0

㊶	㊷	㊸	㊹	㊺	㊻	㊼	㊽	㊾	㊿
1	1	12	4	0	10	0	6	9	11
× 0	× 7	× 2	× 1	× 5	× 6	× 3	×12	×10	× 8

Mixed Facts (0-12)

① 1 × 6 ② 5 × 4 ③ 2 × 8 ④ 2 × 9 ⑤ 7 × 3 ⑥ 0 × 5 ⑦ 6 ×10 ⑧ 5 × 7 ⑨ 0 × 0 ⑩ 12 ×12

⑪ 5 ×11 ⑫ 6 × 1 ⑬ 0 ×10 ⑭ 9 × 2 ⑮ 5 × 6 ⑯ 12 × 5 ⑰ 2 × 0 ⑱ 8 × 9 ⑲ 10 × 3 ⑳ 2 × 4

㉑ 10 × 4 ㉒ 7 × 9 ㉓ 8 × 7 ㉔ 0 ×12 ㉕ 2 ×11 ㉖ 5 × 2 ㉗ 5 × 8 ㉘ 12 × 6 ㉙ 9 × 1 ㉚ 1 × 0

㉛ 9 × 7 ㉜ 0 × 8 ㉝ 2 × 3 ㉞ 1 × 5 ㉟ 6 × 0 ㊱ 8 × 1 ㊲ 10 ×12 ㊳ 7 ×12 ㊴ 5 ×10 ㊵ 0 ×11

㊶ 8 ×12 ㊷ 0 × 6 ㊸ 12 × 2 ㊹ 7 ×10 ㊺ 9 × 4 ㊻ 2 × 5 ㊼ 2 × 3 ㊽ 5 ×11 ㊾ 1 × 1 ㊿ 6 × 9

Mixed Facts (0-12)

① 12 ×10 ② 1 × 3 ③ 5 × 0 ④ 5 × 5 ⑤ 10 × 9 ⑥ 2 × 6 ⑦ 7 × 4 ⑧ 0 × 2 ⑨ 0 × 8 ⑩ 2 × 7

⑪ 2 × 2 ⑫ 10 ×11 ⑬ 6 × 4 ⑭ 5 × 1 ⑮ 0 × 7 ⑯ 9 × 8 ⑰ 5 × 5 ⑱ 2 ×12 ⑲ 12 × 0 ⑳ 8 ×10

㉑ 5 × 9 ㉒ 9 × 0 ㉓ 7 × 1 ㉔ 0 × 7 ㉕ 1 ×10 ㉖ 5 ×12 ㉗ 8 ×11 ㉘ 10 × 8 ㉙ 6 × 6 ㉚ 0 × 3

㉛ 0 × 3 ㉜ 12 × 7 ㉝ 0 × 6 ㉞ 8 × 4 ㉟ 2 × 1 ㊱ 6 ×11 ㊲ 2 × 9 ㊳ 1 × 2 ㊴ 7 ×12 ㊵ 9 × 5

㊶ 6 × 5 ㊷ 2 ×10 ㊸ 1 × 9 ㊹ 10 × 2 ㊺ 5 × 6 ㊻ 7 × 0 ㊼ 0 × 1 ㊽ 0 × 4 ㊾ 2 ×11 ㊿ 5 × 8

No Zero or One (2-12)

①   3   ②   11   ③   7   ④   9   ⑤   5   ⑥   10   ⑦   8   ⑧   4   ⑨   12   ⑩   6
× 3   ×12   × 4   × 8   × 7   ×11   ×12   × 2   × 8   × 4

⑪   12   ⑫   4   ⑬   8   ⑭   2   ⑮   11   ⑯   6   ⑰   9   ⑱   7   ⑲   5   ⑳   10
× 9   × 9   × 2   × 4   × 5   × 9   ×11   ×12   × 6   × 7

㉑   9   ㉒   5   ㉓   3   ㉔   6   ㉕   10   ㉖   7   ㉗   2   ㉘   11   ㉙   12   ㉚   8
× 3   ×10   ×12   ×11   × 6   × 8   ×10   × 4   × 3   ×10

㉛   2   ㉜   12   ㉝   10   ㉞   8   ㉟   9   ㊱   4   ㊲   11   ㊳   6   ㊴   7   ㊵   3
× 9   ×10   × 5   × 7   × 5   × 4   × 3   × 8   × 9   × 5

㊶   10   ㊷   9   ㊸   12   ㊹   7   ㊺   3   ㊻   11   ㊼   5   ㊽   8   ㊾   2   ㊿   4
× 2   ×10   ×11   × 5   × 6   × 9   ×12   × 8   × 6   ×10

No Zero or One (2-12)

① 5 × 3
② 6 × 2
③ 2 × 8
④ 4 × 3
⑤ 12 × 2
⑥ 8 × 6
⑦ 10 ×10
⑧ 3 ×11
⑨ 9 × 4
⑩ 11 × 8

⑪ 4 × 5
⑫ 3 × 7
⑬ 12 ×12
⑭ 5 ×11
⑮ 6 × 7
⑯ 2 × 2
⑰ 7 ×11
⑱ 10 × 4
⑲ 11 × 2
⑳ 9 × 2

㉑ 12 × 5
㉒ 8 ×11
㉓ 11 × 6
㉔ 10 × 3
㉕ 7 × 6
㉖ 3 ×10
㉗ 4 ×12
㉘ 5 × 2
㉙ 6 ×12
㉚ 2 × 7

㉛ 11 × 7
㉜ 2 × 3
㉝ 4 ×11
㉞ 12 × 6
㉟ 8 × 3
㊱ 5 × 9
㊲ 6 × 5
㊳ 9 ×12
㊴ 3 × 2
㊵ 7 ×10

㊶ 7 × 7
㊷ 10 ×12
㊸ 6 × 6
㊹ 3 × 4
㊺ 4 × 8
㊻ 9 × 9
㊼ 12 × 7
㊽ 2 × 5
㊾ 8 × 9
㊿ 5 × 8

No Zero or One (2-12)

① 3 ×9
② 6 ×11
③ 8 ×10
④ 7 × 5
⑤ 10 × 8
⑥ 9 × 7
⑦ 4 ×11
⑧ 2 × 3
⑨ 12 × 5
⑩ 11 ×10

⑪ 12 × 2
⑫ 2 × 2
⑬ 4 × 3
⑭ 5 ×10
⑮ 6 × 4
⑯ 11 × 2
⑰ 7 × 7
⑱ 8 ×11
⑲ 10 ×12
⑳ 9 × 8

㉑ 7 × 9
㉒ 10 × 6
㉓ 3 ×11
㉔ 11 × 7
㉕ 9 ×12
㉖ 8 × 5
㉗ 5 × 6
㉘ 6 ×10
㉙ 12 × 9
㉚ 4 × 6

㉛ 5 × 2
㉜ 12 × 6
㉝ 9 × 4
㉞ 4 × 8
㉟ 7 × 4
㊱ 2 ×10
㊲ 6 × 9
㊳ 11 × 5
㊴ 8 × 2
㊵ 3 × 4

㊶ 9 × 3
㊷ 7 × 6
㊸ 12 × 7
㊹ 8 × 4
㊺ 3 ×12
㊻ 6 × 2
㊼ 10 ×11
㊽ 4 × 5
㊾ 5 ×12
㊿ 2 × 6

No Zero or One (2-12)

①
```
   10
×   9
```
②
```
   11
×    3
```
③
```
    5
×   5
```
④
```
    2
×   9
```
⑤
```
   12
×    3
```
⑥
```
    4
×12
```
⑦
```
    9
×   6
```
⑧
```
    3
×   7
```
⑨
```
    7
×10
```
⑩
```
    6
×   5
```

⑪
```
    2
×   4
```
⑫
```
    3
×   8
```
⑬
```
   12
×11
```
⑭
```
   10
×   7
```
⑮
```
   11
×   8
```
⑯
```
    5
×   3
```
⑰
```
    8
×   7
```
⑱
```
    9
×10
```
⑲
```
    6
×   3
```
⑳
```
    7
×   3
```

㉑
```
   12
×   4
```
㉒
```
    4
×   7
```
㉓
```
    6
×12
```
㉔
```
    9
×   9
```
㉕
```
    8
×12
```
㉖
```
    3
×   6
```
㉗
```
    2
×11
```
㉘
```
   10
×   3
```
㉙
```
   11
×11
```
㉚
```
    5
×   8
```

㉛
```
    6
×   8
```
㉜
```
    5
×   9
```
㉝
```
    2
×   7
```
㉞
```
   12
×12
```
㉟
```
    4
×   9
```
㊱
```
   10
×   2
```
㊲
```
   11
×   4
```
㊳
```
    7
×11
```
㊴
```
    3
×   3
```
㊵
```
    8
×   6
```

㊶
```
    8
×   8
```
㊷
```
    9
×11
```
㊸
```
   11
×12
```
㊹
```
    3
×10
```
㊺
```
    2
×   5
```
㊻
```
    7
×   2
```
㊼
```
   12
×   8
```
㊽
```
    5
×   4
```
㊾
```
    4
×   2
```
㊿
```
   10
×   5
```

No Zero or One (2-12)

① 5 × 8
② 2 ×11
③ 3 × 6
④ 6 × 7
⑤ 4 ×12
⑥ 11 × 9
⑦ 8 ×11
⑧ 7 × 4
⑨ 12 × 7
⑩ 10 × 6

⑪ 12 × 2
⑫ 7 × 2
⑬ 8 × 4
⑭ 9 × 6
⑮ 2 × 5
⑯ 10 × 2
⑰ 6 × 9
⑱ 3 ×11
⑲ 4 × 3
⑳ 11 ×12

㉑ 6 × 8
㉒ 4 ×10
㉓ 5 ×11
㉔ 10 × 9
㉕ 11 × 3
㉖ 3 × 7
㉗ 9 ×10
㉘ 2 × 6
㉙ 12 × 8
㉚ 8 ×10

㉛ 9 × 2
㉜ 12 ×10
㉝ 11 × 5
㉞ 8 ×12
㉟ 6 × 5
㊱ 7 × 6
㊲ 2 × 8
㊳ 10 × 7
㊴ 3 × 2
㊵ 5 × 5

㊶ 11 × 4
㊷ 6 ×10
㊸ 12 × 9
㊹ 3 × 5
㊺ 5 × 3
㊻ 2 × 2
㊼ 4 ×11
㊽ 8 × 7
㊾ 9 × 3
㊿ 7 ×10

No Zero or One (2-12)

①
$$\begin{array}{r} 4 \\ \times\ 8 \\ \hline \end{array}$$
②
$$\begin{array}{r} 10 \\ \times\ 4 \\ \hline \end{array}$$
③
$$\begin{array}{r} 9 \\ \times\ 7 \\ \hline \end{array}$$
④
$$\begin{array}{r} 7 \\ \times\ 8 \\ \hline \end{array}$$
⑤
$$\begin{array}{r} 12 \\ \times\ 4 \\ \hline \end{array}$$
⑥
$$\begin{array}{r} 8 \\ \times\ 3 \\ \hline \end{array}$$
⑦
$$\begin{array}{r} 11 \\ \times 10 \\ \hline \end{array}$$
⑧
$$\begin{array}{r} 5 \\ \times\ 9 \\ \hline \end{array}$$
⑨
$$\begin{array}{r} 6 \\ \times\ 6 \\ \hline \end{array}$$
⑩
$$\begin{array}{r} 2 \\ \times\ 7 \\ \hline \end{array}$$

⑪
$$\begin{array}{r} 7 \\ \times\ 5 \\ \hline \end{array}$$
⑫
$$\begin{array}{r} 5 \\ \times 12 \\ \hline \end{array}$$
⑬
$$\begin{array}{r} 12 \\ \times 11 \\ \hline \end{array}$$
⑭
$$\begin{array}{r} 4 \\ \times\ 9 \\ \hline \end{array}$$
⑮
$$\begin{array}{r} 10 \\ \times 12 \\ \hline \end{array}$$
⑯
$$\begin{array}{r} 9 \\ \times\ 4 \\ \hline \end{array}$$
⑰
$$\begin{array}{r} 3 \\ \times\ 9 \\ \hline \end{array}$$
⑱
$$\begin{array}{r} 11 \\ \times\ 6 \\ \hline \end{array}$$
⑲
$$\begin{array}{r} 2 \\ \times\ 4 \\ \hline \end{array}$$
⑳
$$\begin{array}{r} 6 \\ \times\ 4 \\ \hline \end{array}$$

㉑
$$\begin{array}{r} 12 \\ \times\ 5 \\ \hline \end{array}$$
㉒
$$\begin{array}{r} 8 \\ \times\ 9 \\ \hline \end{array}$$
㉓
$$\begin{array}{r} 2 \\ \times\ 3 \\ \hline \end{array}$$
㉔
$$\begin{array}{r} 11 \\ \times\ 8 \\ \hline \end{array}$$
㉕
$$\begin{array}{r} 3 \\ \times\ 3 \\ \hline \end{array}$$
㉖
$$\begin{array}{r} 5 \\ \times 10 \\ \hline \end{array}$$
㉗
$$\begin{array}{r} 7 \\ \times 11 \\ \hline \end{array}$$
㉘
$$\begin{array}{r} 4 \\ \times\ 4 \\ \hline \end{array}$$
㉙
$$\begin{array}{r} 10 \\ \times 11 \\ \hline \end{array}$$
㉚
$$\begin{array}{r} 9 \\ \times 12 \\ \hline \end{array}$$

㉛
$$\begin{array}{r} 2 \\ \times 12 \\ \hline \end{array}$$
㉜
$$\begin{array}{r} 9 \\ \times\ 8 \\ \hline \end{array}$$
㉝
$$\begin{array}{r} 7 \\ \times\ 9 \\ \hline \end{array}$$
㉞
$$\begin{array}{r} 12 \\ \times\ 3 \\ \hline \end{array}$$
㉟
$$\begin{array}{r} 8 \\ \times\ 8 \\ \hline \end{array}$$
㊱
$$\begin{array}{r} 4 \\ \times\ 2 \\ \hline \end{array}$$
㊲
$$\begin{array}{r} 10 \\ \times\ 5 \\ \hline \end{array}$$
㊳
$$\begin{array}{r} 6 \\ \times 11 \\ \hline \end{array}$$
㊴
$$\begin{array}{r} 5 \\ \times\ 4 \\ \hline \end{array}$$
㊵
$$\begin{array}{r} 3 \\ \times 10 \\ \hline \end{array}$$

㊶
$$\begin{array}{r} 3 \\ \times 12 \\ \hline \end{array}$$
㊷
$$\begin{array}{r} 11 \\ \times 11 \\ \hline \end{array}$$
㊸
$$\begin{array}{r} 10 \\ \times\ 3 \\ \hline \end{array}$$
㊹
$$\begin{array}{r} 5 \\ \times\ 6 \\ \hline \end{array}$$
㊺
$$\begin{array}{r} 7 \\ \times\ 7 \\ \hline \end{array}$$
㊻
$$\begin{array}{r} 6 \\ \times\ 2 \\ \hline \end{array}$$
㊼
$$\begin{array}{r} 12 \\ \times 12 \\ \hline \end{array}$$
㊽
$$\begin{array}{r} 9 \\ \times\ 5 \\ \hline \end{array}$$
㊾
$$\begin{array}{r} 8 \\ \times\ 2 \\ \hline \end{array}$$
㊿
$$\begin{array}{r} 4 \\ \times\ 7 \\ \hline \end{array}$$

No Zero or One (2-12)

①
$$\begin{array}{r} 5 \\ \times\ 5 \\ \hline \end{array}$$

②
$$\begin{array}{r} 3 \\ \times\ 2 \\ \hline \end{array}$$

③
$$\begin{array}{r} 2 \\ \times\ 4 \\ \hline \end{array}$$

④
$$\begin{array}{r} 6 \\ \times\ 6 \\ \hline \end{array}$$

⑤
$$\begin{array}{r} 10 \\ \times 12 \\ \hline \end{array}$$

⑥
$$\begin{array}{r} 11 \\ \times\ 7 \\ \hline \end{array}$$

⑦
$$\begin{array}{r} 7 \\ \times\ 2 \\ \hline \end{array}$$

⑧
$$\begin{array}{r} 9 \\ \times\ 8 \\ \hline \end{array}$$

⑨
$$\begin{array}{r} 8 \\ \times\ 6 \\ \hline \end{array}$$

⑩
$$\begin{array}{r} 4 \\ \times\ 4 \\ \hline \end{array}$$

⑪
$$\begin{array}{r} 8 \\ \times 11 \\ \hline \end{array}$$

⑫
$$\begin{array}{r} 9 \\ \times 11 \\ \hline \end{array}$$

⑬
$$\begin{array}{r} 7 \\ \times\ 8 \\ \hline \end{array}$$

⑭
$$\begin{array}{r} 12 \\ \times\ 4 \\ \hline \end{array}$$

⑮
$$\begin{array}{r} 3 \\ \times 10 \\ \hline \end{array}$$

⑯
$$\begin{array}{r} 4 \\ \times 11 \\ \hline \end{array}$$

⑰
$$\begin{array}{r} 6 \\ \times\ 7 \\ \hline \end{array}$$

⑱
$$\begin{array}{r} 2 \\ \times\ 2 \\ \hline \end{array}$$

⑲
$$\begin{array}{r} 10 \\ \times\ 9 \\ \hline \end{array}$$

⑳
$$\begin{array}{r} 11 \\ \times 12 \\ \hline \end{array}$$

㉑
$$\begin{array}{r} 6 \\ \times\ 5 \\ \hline \end{array}$$

㉒
$$\begin{array}{r} 10 \\ \times\ 3 \\ \hline \end{array}$$

㉓
$$\begin{array}{r} 5 \\ \times\ 2 \\ \hline \end{array}$$

㉔
$$\begin{array}{r} 4 \\ \times\ 7 \\ \hline \end{array}$$

㉕
$$\begin{array}{r} 11 \\ \times\ 9 \\ \hline \end{array}$$

㉖
$$\begin{array}{r} 2 \\ \times\ 6 \\ \hline \end{array}$$

㉗
$$\begin{array}{r} 12 \\ \times\ 3 \\ \hline \end{array}$$

㉘
$$\begin{array}{r} 3 \\ \times\ 4 \\ \hline \end{array}$$

㉙
$$\begin{array}{r} 8 \\ \times\ 5 \\ \hline \end{array}$$

㉚
$$\begin{array}{r} 7 \\ \times\ 3 \\ \hline \end{array}$$

㉛
$$\begin{array}{r} 12 \\ \times 11 \\ \hline \end{array}$$

㉜
$$\begin{array}{r} 8 \\ \times\ 3 \\ \hline \end{array}$$

㉝
$$\begin{array}{r} 11 \\ \times 10 \\ \hline \end{array}$$

㉞
$$\begin{array}{r} 7 \\ \times 12 \\ \hline \end{array}$$

㉟
$$\begin{array}{r} 6 \\ \times 10 \\ \hline \end{array}$$

㊱
$$\begin{array}{r} 9 \\ \times\ 4 \\ \hline \end{array}$$

㊲
$$\begin{array}{r} 3 \\ \times\ 5 \\ \hline \end{array}$$

㊳
$$\begin{array}{r} 4 \\ \times\ 6 \\ \hline \end{array}$$

㊴
$$\begin{array}{r} 2 \\ \times 11 \\ \hline \end{array}$$

㊵
$$\begin{array}{r} 5 \\ \times 10 \\ \hline \end{array}$$

㊶
$$\begin{array}{r} 11 \\ \times\ 8 \\ \hline \end{array}$$

㊷
$$\begin{array}{r} 6 \\ \times\ 3 \\ \hline \end{array}$$

㊸
$$\begin{array}{r} 8 \\ \times\ 7 \\ \hline \end{array}$$

㊹
$$\begin{array}{r} 2 \\ \times 10 \\ \hline \end{array}$$

㊺
$$\begin{array}{r} 5 \\ \times\ 9 \\ \hline \end{array}$$

㊻
$$\begin{array}{r} 3 \\ \times 11 \\ \hline \end{array}$$

㊼
$$\begin{array}{r} 10 \\ \times\ 2 \\ \hline \end{array}$$

㊽
$$\begin{array}{r} 7 \\ \times\ 6 \\ \hline \end{array}$$

㊾
$$\begin{array}{r} 12 \\ \times\ 9 \\ \hline \end{array}$$

㊿
$$\begin{array}{r} 9 \\ \times\ 3 \\ \hline \end{array}$$

No Zero or One (2-12)

① 10 × 5 ② 4 × 8 ③ 12 × 6 ④ 9 × 5 ⑤ 8 × 8 ⑥ 7 × 9 ⑦ 11 × 3 ⑧ 5 × 7 ⑨ 6 × 4 ⑩ 3 × 6

⑪ 9 ×10 ⑫ 5 ×12 ⑬ 8 × 2 ⑭ 10 × 7 ⑮ 4 ×12 ⑯ 12 × 8 ⑰ 2 × 7 ⑱ 11 × 4 ⑲ 3 × 8 ⑳ 6 × 8

㉑ 8 ×10 ㉒ 7 × 7 ㉓ 3 × 9 ㉔ 11 × 5 ㉕ 2 × 9 ㉖ 5 × 3 ㉗ 9 × 2 ㉘ 10 × 8 ㉙ 4 × 2 ㉚ 12 ×12

㉛ 3 ×12 ㉜ 12 × 5 ㉝ 9 × 7 ㉞ 8 × 9 ㉟ 7 × 5 ㊱ 10 ×11 ㊲ 4 ×10 ㊳ 6 × 2 ㊴ 5 × 8 ㊵ 2 × 3

㊶ 2 ×12 ㊷ 11 × 2 ㊸ 4 × 9 ㊹ 5 × 4 ㊺ 9 × 6 ㊻ 6 ×11 ㊼ 8 ×12 ㊽ 12 ×10 ㊾ 7 ×11 ㊿ 10 × 6

No Zero or One (2-12)

①	②	③	④	⑤	⑥	⑦	⑧	⑨	⑩
6	8	5	12	7	3	10	9	11	2
× 6	× 3	×12	× 7	× 4	×11	× 3	×10	× 7	×12

⑪	⑫	⑬	⑭	⑮	⑯	⑰	⑱	⑲	⑳
11	9	10	4	8	2	12	5	7	3
× 8	× 8	×10	×12	× 9	× 8	×11	× 3	× 2	× 4

㉑	㉒	㉓	㉔	㉕	㉖	㉗	㉘	㉙	㉚
12	7	6	2	3	5	4	8	11	10
× 6	× 5	× 3	×11	× 2	× 7	× 5	×12	× 6	× 5

㉛	㉜	㉝	㉞	㉟	㊱	㊲	㊳	㊴	㊵
4	11	3	10	12	9	8	2	5	6
× 8	× 5	× 9	× 4	× 9	×12	× 6	× 7	× 8	× 9

㊶	㊷	㊸	㊹	㊺	㊻	㊼	㊽	㊾	㊿
3	12	11	5	6	8	7	10	4	9
×10	× 5	×11	× 9	× 2	× 8	× 3	× 7	× 2	× 5

No Zero or One (2-12)

①	②	③	④	⑤	⑥	⑦	⑧	⑨	⑩
7	2	4	9	11	10	3	6	12	8
× 6	×10	× 7	× 6	×10	× 2	× 5	×11	×12	× 7

⑪	⑫	⑬	⑭	⑮	⑯	⑰	⑱	⑲	⑳
9	6	11	7	2	4	5	3	8	12
× 9	× 4	× 3	×11	× 4	×10	×11	×12	×10	×10

㉑	㉒	㉓	㉔	㉕	㉖	㉗	㉘	㉙	㉚
11	10	8	3	5	6	9	7	2	4
× 9	×11	× 2	× 6	× 2	× 5	× 3	×10	× 3	× 4

㉛	㉜	㉝	㉞	㉟	㊱	㊲	㊳	㊴	㊵
8	4	9	11	10	7	2	12	6	5
× 4	× 6	×11	× 2	× 6	× 8	× 9	× 3	×10	× 5

㊶	㊷	㊸	㊹	㊺	㊻	㊼	㊽	㊾	㊿
5	3	2	6	9	12	11	4	10	7
× 4	× 3	× 2	×12	× 7	× 8	× 4	× 9	× 8	× 7

ANSWER KEY

Page 11

① 0	② 0	③ 0	④ 0	⑤ 0	⑥ 0	⑦ 0	⑧ 0	⑨ 0	⑩ 0
⑪ 0	⑫ 0	⑬ 0	⑭ 0	⑮ 0	⑯ 0	⑰ 0	⑱ 0	⑲ 0	⑳ 0
㉑ 0	㉒ 0	㉓ 0	㉔ 0	㉕ 0	㉖ 0	㉗ 0	㉘ 0	㉙ 0	㉚ 0
㉛ 0	㉜ 0	㉝ 0	㉞ 0	㉟ 0	㊱ 0	㊲ 0	㊳ 0	㊴ 0	㊵ 0
㊶ 0	㊷ 0	㊸ 0	㊹ 0	㊺ 0	㊻ 0	㊼ 0	㊽ 0	㊾ 0	㊿ 0

Page 13

① 7	② 4	③ 3	④ 5	⑤ 0	⑥ 2	⑦ 6	⑧ 8	⑨ 1	⑩ 9
⑪ 0	⑫ 6	⑬ 2	⑭ 8	⑮ 4	⑯ 1	⑰ 9	⑱ 7	⑲ 5	⑳ 3
㉑ 8	㉒ 5	㉓ 9	㉔ 6	㉕ 7	㉖ 3	㉗ 1	㉘ 0	㉙ 2	㉚ 4
㉛ 4	㉜ 7	㉝ 1	㉞ 9	㉟ 2	㊱ 0	㊲ 8	㊳ 5	㊴ 3	㊵ 6
㊶ 5	㊷ 3	㊸ 6	㊹ 0	㊺ 9	㊻ 8	㊼ 4	㊽ 2	㊾ 7	㊿ 1

Page 15

① 18	② 6	③ 12	④ 8	⑤ 14	⑥ 4	⑦ 0	⑧ 16	⑨ 10	⑩ 2
⑪ 14	⑫ 0	⑬ 4	⑭ 16	⑮ 6	⑯ 10	⑰ 2	⑱ 18	⑲ 8	⑳ 12
㉑ 16	㉒ 8	㉓ 2	㉔ 0	㉕ 18	㉖ 12	㉗ 10	㉘ 14	㉙ 4	㉚ 6
㉛ 6	㉜ 18	㉝ 10	㉞ 2	㉟ 4	㊱ 14	㊲ 16	㊳ 8	㊴ 12	㊵ 0
㊶ 8	㊷ 12	㊸ 0	㊹ 14	㊺ 2	㊻ 16	㊼ 6	㊽ 4	㊾ 18	㊿ 10

Page 17

① 9	② 15	③ 27	④ 0	⑤ 24	⑥ 18	⑦ 3	⑧ 12	⑨ 6	⑩ 21
⑪ 24	⑫ 3	⑬ 18	⑭ 12	⑮ 15	⑯ 6	⑰ 21	⑱ 9	⑲ 0	⑳ 27
㉑ 12	㉒ 0	㉓ 21	㉔ 3	㉕ 9	㉖ 27	㉗ 6	㉘ 24	㉙ 18	㉚ 15
㉛ 15	㉜ 9	㉝ 6	㉞ 21	㉟ 18	㊱ 24	㊲ 12	㊳ 0	㊴ 27	㊵ 3
㊶ 0	㊷ 27	㊸ 3	㊹ 24	㊺ 21	㊻ 12	㊼ 15	㊽ 18	㊾ 9	㊿ 6

Page 19

① 0	② 32	③ 24	④ 28	⑤ 16	⑥ 4	⑦ 20	⑧ 36	⑨ 8	⑩ 12
⑪ 16	⑫ 20	⑬ 4	⑭ 36	⑮ 32	⑯ 8	⑰ 12	⑱ 0	⑲ 28	⑳ 24
㉑ 36	㉒ 28	㉓ 12	㉔ 20	㉕ 0	㉖ 24	㉗ 8	㉘ 16	㉙ 4	㉚ 32
㉛ 32	㉜ 0	㉝ 8	㉞ 12	㉟ 4	㊱ 16	㊲ 36	㊳ 28	㊴ 24	㊵ 20
㊶ 28	㊷ 24	㊸ 20	㊹ 16	㊺ 12	㊻ 36	㊼ 32	㊽ 4	㊾ 0	㊿ 8

Page 21

① 40	② 10	③ 30	④ 45	⑤ 35	⑥ 20	⑦ 15	⑧ 25	⑨ 0	⑩ 5
⑪ 35	⑫ 15	⑬ 20	⑭ 25	⑮ 10	⑯ 0	⑰ 5	⑱ 40	⑲ 45	⑳ 30
㉑ 25	㉒ 45	㉓ 5	㉔ 15	㉕ 40	㉖ 30	㉗ 0	㉘ 35	㉙ 20	㉚ 10
㉛ 10	㉜ 40	㉝ 0	㉞ 5	㉟ 20	㊱ 35	㊲ 25	㊳ 45	㊴ 30	㊵ 15
㊶ 45	㊷ 30	㊸ 15	㊹ 35	㊺ 5	㊻ 25	㊼ 10	㊽ 20	㊾ 40	㊿ 0

Page 23

① 6	② 42	③ 36	④ 54	⑤ 24	⑥ 48	⑦ 30	⑧ 18	⑨ 12	⑩ 0
⑪ 24	⑫ 30	⑬ 48	⑭ 18	⑮ 42	⑯ 12	⑰ 0	⑱ 6	⑲ 54	⑳ 36
㉑ 18	㉒ 54	㉓ 0	㉔ 30	㉕ 6	㉖ 36	㉗ 12	㉘ 24	㉙ 48	㉚ 42
㉛ 42	㉜ 6	㉝ 12	㉞ 0	㉟ 48	㊱ 24	㊲ 18	㊳ 54	㊴ 36	㊵ 30
㊶ 54	㊷ 36	㊸ 30	㊹ 24	㊺ 0	㊻ 18	㊼ 42	㊽ 48	㊾ 6	㊿ 12

Page 25

① 14	② 35	③ 21	④ 0	⑤ 28	⑥ 42	⑦ 49	⑧ 56	⑨ 63	⑩ 7
⑪ 28	⑫ 49	⑬ 42	⑭ 56	⑮ 35	⑯ 63	⑰ 7	⑱ 14	⑲ 0	⑳ 21
㉑ 56	㉒ 0	㉓ 7	㉔ 49	㉕ 14	㉖ 21	㉗ 63	㉘ 28	㉙ 42	㉚ 35
㉛ 35	㉜ 14	㉝ 63	㉞ 7	㉟ 42	㊱ 28	㊲ 56	㊳ 0	㊴ 21	㊵ 49
㊶ 0	㊷ 21	㊸ 49	㊹ 28	㊺ 7	㊻ 56	㊼ 35	㊽ 42	㊾ 14	㊿ 63

Page 27

① 40	② 48	③ 16	④ 72	⑤ 24	⑥ 32	⑦ 8	⑧ 0	⑨ 56	⑩ 64
⑪ 24	⑫ 8	⑬ 32	⑭ 0	⑮ 48	⑯ 56	⑰ 64	⑱ 40	⑲ 72	⑳ 16
㉑ 0	㉒ 72	㉓ 64	㉔ 8	㉕ 40	㉖ 16	㉗ 56	㉘ 24	㉙ 32	㉚ 48
㉛ 48	㉜ 40	㉝ 56	㉞ 64	㉟ 32	㊱ 24	㊲ 0	㊳ 72	㊴ 16	㊵ 8
㊶ 72	㊷ 16	㊸ 8	㊹ 24	㊺ 64	㊻ 0	㊼ 48	㊽ 32	㊾ 40	㊿ 56

Page 29

①54	②81	③0	④18	⑤72	⑥45	⑦63	⑧27	⑨36	⑩9
⑪72	⑫63	⑬45	⑭27	⑮81	⑯36	⑰9	⑱54	⑲18	⑳0
㉑27	㉒18	㉓9	㉔63	㉕54	㉖0	㉗36	㉘72	㉙45	㉚81
㉛81	㉜54	㉝36	㉞9	㉟45	㊱72	㊲27	㊳18	㊴0	㊵63
㊶18	㊷0	㊸63	㊹72	㊺9	㊻27	㊼81	㊽45	㊾54	㊿36

Page 30

①10	②10	③0	④3	⑤0	⑥0	⑦9	⑧4	⑨3	⑩0
⑪0	⑫9	⑬0	⑭1	⑮20	⑯20	⑰0	⑱4	⑲20	⑳4
㉑5	㉒16	㉓3	㉔25	㉕6	㉖8	㉗2	㉘0	㉙0	㉚12
㉛8	㉜5	㉝3	㉞0	㉟8	㊱0	㊲0	㊳15	㊴10	㊵12
㊶12	㊷0	㊸8	㊹6	㊺0	㊻5	㊼2	㊽0	㊾15	㊿4

Page 31

①5	②0	③2	④0	⑤15	⑥0	⑦8	⑧15	⑨5	⑩3
⑪3	⑫20	⑬20	⑭6	⑮0	⑯12	⑰0	⑱12	⑲4	⑳0
㉑8	㉒0	㉓9	㉔0	㉕20	㉖3	㉗10	㉘1	㉙0	㉚12
㉛4	㉜10	㉝20	㉞4	㉟0	㊱0	㊲6	㊳0	㊴0	㊵8
㊶0	㊷3	㊸0	㊹25	㊺2	㊻16	㊼0	㊽6	㊾10	㊿3

Page 32

①20	②6	③0	④10	⑤1	⑥0	⑦6	⑧10	⑨3	⑩20
⑪4	⑫6	⑬5	⑭5	⑮0	⑯0	⑰12	⑱0	⑲0	⑳10
㉑15	㉒0	㉓3	㉔12	㉕4	㉖0	㉗2	㉘2	㉙16	㉚0
㉛0	㉜4	㉝10	㉞4	㉟0	㊱12	㊲0	㊳9	㊴20	㊵0
㊶0	㊷4	㊸0	㊹15	㊺3	㊻4	㊼25	㊽8	㊾8	㊿0

Page 33

①15	②5	③2	④0	⑤8	⑥5	⑦0	⑧9	⑨4	⑩10
⑪10	⑫0	⑬0	⑭15	⑮1	⑯0	⑰20	⑱0	⑲10	⑳3
㉑0	㉒8	㉓6	㉔0	㉕0	㉖10	㉗20	㉘5	㉙4	㉚0
㉛0	㉜6	㉝0	㉞0	㉟12	㊱3	㊲4	㊳16	㊴2	㊵0
㊶4	㊷10	㊸8	㊹12	㊺25	㊻0	㊼1	㊽15	㊾6	㊿3

Page 34

① 1	② 16	③ 15	④ 0	⑤ 4	⑥ 0	⑦ 15	⑧ 4	⑨ 6	⑩ 0
⑪ 6	⑫ 15	⑬ 2	⑭ 0	⑮ 0	⑯ 20	⑰ 9	⑱ 0	⑲ 20	⑳ 4
㉑ 0	㉒ 0	㉓ 6	㉔ 4	㉕ 20	㉖ 5	㉗ 8	㉘ 10	㉙ 3	㉚ 0
㉛ 0	㉜ 2	㉝ 0	㉞ 6	㉟ 5	㊱ 9	㊲ 0	㊳ 12	㊴ 1	㊵ 25
㊶ 25	㊷ 6	㊸ 0	㊹ 3	㊺ 8	㊻ 2	㊼ 0	㊽ 12	㊾ 5	㊿ 10

Page 35

① 0	② 2	③ 8	④ 15	⑤ 5	⑥ 2	⑦ 0	⑧ 12	⑨ 2	⑩ 0
⑪ 0	⑫ 0	⑬ 20	⑭ 3	⑮ 4	⑯ 0	⑰ 0	⑱ 25	⑲ 4	⑳ 8
㉑ 5	㉒ 12	㉓ 15	㉔ 0	㉕ 20	㉖ 0	㉗ 1	㉘ 0	㉙ 6	㉚ 0
㉛ 0	㉜ 16	㉝ 0	㉞ 10	㉟ 9	㊱ 8	㊲ 20	㊳ 3	㊴ 10	㊵ 5
㊶ 6	㊷ 0	㊸ 12	㊹ 4	㊺ 0	㊻ 0	㊼ 4	㊽ 3	㊾ 16	㊿ 6

Page 36

① 15	② 0	③ 4	④ 0	⑤ 4	⑥ 2	⑦ 15	⑧ 12	⑨ 6	⑩ 0
⑪ 8	⑫ 15	⑬ 6	⑭ 0	⑮ 5	⑯ 0	⑰ 12	⑱ 0	⑲ 0	⑳ 12
㉑ 0	㉒ 1	㉓ 6	㉔ 0	㉕ 20	㉖ 3	㉗ 8	㉘ 10	㉙ 20	㉚ 3
㉛ 4	㉜ 10	㉝ 0	㉞ 8	㉟ 3	㊱ 12	㊲ 2	㊳ 0	㊴ 15	㊵ 5
㊶ 5	㊷ 8	㊸ 4	㊹ 9	㊺ 0	㊻ 10	㊼ 0	㊽ 16	㊾ 25	㊿ 2

Page 37

① 0	② 6	③ 8	④ 4	⑤ 25	⑥ 6	⑦ 4	⑧ 0	⑨ 10	⑩ 0
⑪ 0	⑫ 5	⑬ 0	⑭ 9	⑮ 4	⑯ 3	⑰ 0	⑱ 5	⑲ 12	⑳ 0
㉑ 3	㉒ 16	㉓ 15	㉔ 2	㉕ 0	㉖ 0	㉗ 15	㉘ 0	㉙ 8	㉚ 3
㉛ 0	㉜ 0	㉝ 5	㉞ 2	㉟ 12	㊱ 0	㊲ 20	㊳ 20	㊴ 10	㊵ 3
㊶ 8	㊷ 0	㊸ 16	㊹ 0	㊺ 0	㊻ 1	㊼ 4	㊽ 9	㊾ 0	㊿ 6

Page 38

① 0	② 12	③ 3	④ 10	⑤ 10	⑥ 20	⑦ 5	⑧ 0	⑨ 2	⑩ 2
⑪ 5	⑫ 5	⑬ 0	⑭ 4	⑮ 0	⑯ 9	⑰ 1	⑱ 8	⑲ 9	⑳ 0
㉑ 6	㉒ 12	㉓ 2	㉔ 0	㉕ 20	㉖ 0	㉗ 8	㉘ 25	㉙ 0	㉚ 4
㉛ 16	㉜ 0	㉝ 10	㉞ 5	㉟ 0	㊱ 1	㊲ 20	㊳ 3	㊴ 0	㊵ 15
㊶ 15	㊷ 5	㊸ 16	㊹ 0	㊺ 15	㊻ 0	㊼ 0	㊽ 4	㊾ 0	㊿ 6

Page 39

① 6	② 0	③ 8	④ 3	⑤ 0	⑥ 0	⑦ 16	⑧ 3	⑨ 0	⑩ 10
⑪ 10	⑫ 0	⑬ 9	⑭ 0	⑮ 10	⑯ 4	⑰ 2	⑱ 15	⑲ 0	⑳ 15
㉑ 0	㉒ 4	㉓ 5	㉔ 20	㉕ 9	㉖ 10	㉗ 0	㉘ 4	㉙ 5	㉚ 4
㉛ 8	㉜ 12	㉝ 0	㉞ 6	㉟ 1	㊱ 15	㊲ 20	㊳ 0	㊴ 25	㊵ 0
㊶ 5	㊷ 10	㊸ 4	㊹ 0	㊺ 0	㊻ 12	㊼ 10	㊽ 0	㊾ 12	㊿ 2

Page 40

① 8	② 0	③ 12	④ 30	⑤ 20	⑥ 0	⑦ 7	⑧ 24	⑨ 8	⑩ 10
⑪ 27	⑫ 20	⑬ 7	⑭ 0	⑮ 14	⑯ 5	⑰ 18	⑱ 0	⑲ 40	⑳ 20
㉑ 8	㉒ 21	㉓ 0	㉔ 32	㉕ 0	㉖ 18	㉗ 20	㉘ 24	㉙ 12	㉚ 9
㉛ 27	㉜ 30	㉝ 4	㉞ 16	㉟ 20	㊱ 16	㊲ 16	㊳ 25	㊴ 28	㊵ 0
㊶ 24	㊷ 18	㊸ 40	㊹ 21	㊺ 45	㊻ 12	㊼ 0	㊽ 5	㊾ 15	㊿ 20

Page 41

① 0	② 4	③ 12	④ 16	⑤ 0	⑥ 32	⑦ 10	⑧ 24	⑨ 36	⑩ 30
⑪ 10	⑫ 27	⑬ 25	⑭ 0	⑮ 4	⑯ 21	⑰ 16	⑱ 7	⑲ 12	⑳ 0
㉑ 18	㉒ 0	㉓ 20	㉔ 18	㉕ 18	㉖ 18	㉗ 0	㉘ 25	㉙ 4	㉚ 21
㉛ 6	㉜ 20	㉝ 21	㉞ 20	㉟ 0	㊱ 8	㊲ 20	㊳ 18	㊴ 24	㊵ 20
㊶ 9	㊷ 35	㊸ 18	㊹ 5	㊺ 32	㊻ 35	㊼ 12	㊽ 36	㊾ 0	㊿ 30

Page 42

① 32	② 25	③ 21	④ 0	⑤ 18	⑥ 25	⑦ 16	⑧ 12	⑨ 32	⑩ 9
⑪ 0	⑫ 18	⑬ 16	⑭ 30	⑮ 4	⑯ 36	⑰ 5	⑱ 35	⑲ 0	⑳ 15
㉑ 7	㉒ 12	㉓ 45	㉔ 16	㉕ 20	㉖ 18	㉗ 0	㉘ 24	㉙ 6	㉚ 20
㉛ 15	㉜ 0	㉝ 28	㉞ 8	㉟ 18	㊱ 14	㊲ 8	㊳ 0	㊴ 8	㊵ 20
㊶ 12	㊷ 5	㊸ 0	㊹ 12	㊺ 0	㊻ 6	㊼ 20	㊽ 36	㊾ 27	㊿ 0

Page 43

① 20	② 28	③ 6	④ 14	⑤ 25	⑥ 16	⑦ 9	⑧ 24	⑨ 10	⑩ 0
⑪ 9	⑫ 15	⑬ 0	⑭ 25	⑮ 28	⑯ 12	⑰ 14	⑱ 16	⑲ 6	⑳ 40
㉑ 18	㉒ 40	㉓ 18	㉔ 5	㉕ 18	㉖ 5	㉗ 40	㉘ 0	㉙ 28	㉚ 12
㉛ 0	㉜ 18	㉝ 12	㉞ 0	㉟ 45	㊱ 32	㊲ 24	㊳ 5	㊴ 24	㊵ 18
㊶ 20	㊷ 0	㊸ 5	㊹ 36	㊺ 16	㊻ 0	㊼ 21	㊽ 10	㊾ 30	㊿ 0

Page 44

① 14 ② 20 ③ 27 ④ 7 ⑤ 16 ⑥ 20 ⑦ 24 ⑧ 14 ⑨ 20 ⑩ 0
⑪ 9 ⑫ 16 ⑬ 24 ⑭ 35 ⑮ 0 ⑯ 32 ⑰ 0 ⑱ 45 ⑲ 5 ⑳ 12
㉑ 0 ㉒ 18 ㉓ 40 ㉔ 10 ㉕ 30 ㉖ 21 ㉗ 9 ㉘ 15 ㉙ 0 ㉚ 16
㉛ 12 ㉜ 7 ㉝ 36 ㉞ 0 ㉟ 16 ㊱ 18 ㊲ 0 ㊳ 8 ㊴ 12 ㊵ 30
㊶ 20 ㊷ 0 ㊸ 5 ㊹ 18 ㊺ 4 ㊻ 0 ㊼ 30 ㊽ 32 ㊾ 24 ㊿ 9

Page 45

① 30 ② 36 ③ 0 ④ 18 ⑤ 20 ⑥ 10 ⑦ 0 ⑧ 15 ⑨ 8 ⑩ 7
⑪ 0 ⑫ 12 ⑬ 8 ⑭ 20 ⑮ 36 ⑯ 18 ⑰ 18 ⑱ 24 ⑲ 0 ⑳ 25
㉑ 21 ㉒ 25 ㉓ 16 ㉔ 0 ㉕ 21 ㉖ 0 ㉗ 25 ㉘ 8 ㉙ 36 ㉚ 18
㉛ 9 ㉜ 16 ㉝ 18 ㉞ 9 ㉟ 40 ㊱ 20 ㊲ 28 ㊳ 0 ㊴ 15 ㊵ 16
㊶ 16 ㊷ 6 ㊸ 0 ㊹ 32 ㊺ 10 ㊻ 6 ㊼ 27 ㊽ 8 ㊾ 35 ㊿ 7

Page 46

① 0 ② 45 ③ 8 ④ 18 ⑤ 10 ⑥ 45 ⑦ 0 ⑧ 12 ⑨ 0 ⑩ 20
⑪ 24 ⑫ 10 ⑬ 0 ⑭ 30 ⑮ 16 ⑯ 0 ⑰ 36 ⑱ 40 ⑲ 21 ⑳ 9
㉑ 32 ㉒ 4 ㉓ 25 ㉔ 14 ㉕ 20 ㉖ 6 ㉗ 24 ㉘ 7 ㉙ 24 ㉚ 0
㉛ 9 ㉜ 18 ㉝ 0 ㉞ 28 ㉟ 10 ㊱ 16 ㊲ 28 ㊳ 15 ㊴ 8 ㊵ 20
㊶ 12 ㊷ 36 ㊸ 21 ㊹ 4 ㊺ 27 ㊻ 24 ㊼ 20 ㊽ 0 ㊾ 5 ㊿ 24

Page 47

① 20 ② 0 ③ 24 ④ 16 ⑤ 45 ⑥ 14 ⑦ 20 ⑧ 7 ⑨ 18 ⑩ 18
⑪ 20 ⑫ 9 ⑬ 15 ⑭ 45 ⑮ 0 ⑯ 4 ⑰ 16 ⑱ 0 ⑲ 24 ⑳ 35
㉑ 6 ㉒ 35 ㉓ 10 ㉔ 36 ㉕ 6 ㉖ 36 ㉗ 35 ㉘ 15 ㉙ 0 ㉚ 4
㉛ 24 ㉜ 10 ㉝ 4 ㉞ 24 ㉟ 25 ㊱ 0 ㊲ 0 ㊳ 36 ㊴ 7 ㊵ 10
㊶ 0 ㊷ 12 ㊸ 36 ㊹ 0 ㊺ 14 ㊻ 12 ㊼ 8 ㊽ 18 ㊾ 30 ㊿ 18

Page 48

① 0 ② 21 ③ 24 ④ 16 ⑤ 9 ⑥ 21 ⑦ 0 ⑧ 8 ⑨ 0 ⑩ 45
⑪ 12 ⑫ 9 ⑬ 0 ⑭ 24 ⑮ 20 ⑯ 0 ⑰ 35 ⑱ 18 ⑲ 10 ⑳ 28
㉑ 30 ㉒ 16 ㉓ 27 ㉔ 5 ㉕ 12 ㉖ 32 ㉗ 12 ㉘ 20 ㉙ 40 ㉚ 0
㉛ 28 ㉜ 16 ㉝ 0 ㉞ 25 ㉟ 9 ㊱ 6 ㊲ 25 ㊳ 18 ㊴ 4 ㊵ 12
㊶ 8 ㊷ 35 ㊸ 10 ㊹ 16 ㊺ 14 ㊻ 40 ㊼ 12 ㊽ 0 ㊾ 36 ㊿ 12

Page 49
①12 ② 0 ③40 ④ 6 ⑤21 ⑥ 5 ⑦45 ⑧20 ⑨ 7 ⑩16
⑪45 ⑫28 ⑬18 ⑭21 ⑮ 0 ⑯16 ⑰ 6 ⑱ 0 ⑲40 ⑳15
㉑32 ㉒15 ㉓ 9 ㉔35 ㉕32 ㉖35 ㉗15 ㉘18 ㉙ 0 ㉚16
㉛12 ㉜ 9 ㉝16 ㉞12 ㉟27 ㊱ 0 ㊲ 0 ㊳35 ㊴20 ㊵ 9
㊶ 0 ㊷ 8 ㊸35 ㊹ 0 ㊺ 5 ㊻ 8 ㊼24 ㊽ 7 ㊾24 ㊿16

Page 50
①45 ②24 ③28 ④72 ⑤48 ⑥42 ⑦45 ⑧20 ⑨40 ⑩32
⑪24 ⑫45 ⑬30 ⑭64 ⑮63 ⑯42 ⑰20 ⑱56 ⑲42 ⑳20
㉑48 ㉒49 ㉓40 ㉔54 ㉕36 ㉖35 ㉗32 ㉘54 ㉙36 ㉚35
㉛28 ㉜72 ㉝72 ㉞24 ㉟35 ㊱20 ㊲42 ㊳30 ㊴45 ㊵63
㊶63 ㊷24 ㊸28 ㊹25 ㊺36 ㊻72 ㊼40 ㊽16 ㊾81 ㊿56

Page 51
①48 ②30 ③32 ④28 ⑤81 ⑥30 ⑦28 ⑧30 ⑨72 ⑩72
⑪72 ⑫63 ⑬42 ⑭25 ⑮48 ⑯35 ⑰32 ⑱63 ⑲20 ⑳36
㉑35 ㉒16 ㉓45 ㉔42 ㉕42 ㉖72 ㉗45 ㉘64 ㉙24 ㉚35
㉛56 ㉜24 ㉝63 ㉞56 ㉟20 ㊱36 ㊲36 ㊳36 ㊴54 ㊵35
㊶24 ㊷72 ㊸16 ㊹54 ㊺40 ㊻49 ㊼48 ㊽25 ㊾24 ㊿40

Page 52
①40 ②63 ③72 ④20 ⑤28 ⑥42 ⑦30 ⑧72 ⑨24 ⑩36
⑪63 ⑫30 ⑬56 ⑭16 ⑮30 ⑯56 ⑰54 ⑱24 ⑲56 ⑳72
㉑28 ㉒48 ㉓24 ㉔35 ㉕45 ㉖64 ㉗36 ㉘35 ㉙45 ㉚36
㉛54 ㉜20 ㉝20 ㉞63 ㉟64 ㊱54 ㊲42 ㊳42 ㊴40 ㊵40
㊶40 ㊷63 ㊸54 ㊹48 ㊺49 ㊻20 ㊼32 ㊽81 ㊾25 ㊿32

Page 53
①28 ②56 ③36 ④72 ⑤25 ⑥56 ⑦54 ⑧42 ⑨20 ⑩20
⑪20 ⑫30 ⑬56 ⑭48 ⑮28 ⑯36 ⑰36 ⑱40 ⑲72 ⑳49
㉑64 ㉒81 ㉓30 ㉔42 ㉕56 ㉖20 ㉗40 ㉘16 ㉙63 ㉚36
㉛24 ㉜63 ㉝30 ㉞32 ㉟54 ㊱49 ㊲45 ㊳45 ㊴35 ㊵64
㊶63 ㊷20 ㊸81 ㊹35 ㊺32 ㊻48 ㊼28 ㊽48 ㊾63 ㊿24

Page 54

① 28 ② 72 ③ 48 ④ 35 ⑤ 20 ⑥ 30 ⑦ 45 ⑧ 56 ⑨ 36 ⑩ 56
⑪ 40 ⑫ 45 ⑬ 35 ⑭ 28 ⑮ 24 ⑯ 54 ⑰ 72 ⑱ 42 ⑲ 54 ⑳ 56
㉑ 63 ㉒ 36 ㉓ 36 ㉔ 36 ㉕ 40 ㉖ 42 ㉗ 32 ㉘ 25 ㉙ 32 ㉚ 54
㉛ 48 ㉜ 16 ㉝ 35 ㉞ 40 ㉟ 42 ㊱ 72 ㊲ 30 ㊳ 81 ㊴ 28 ㊵ 30
㊶ 30 ㊷ 40 ㊸ 48 ㊹ 63 ㊺ 45 ㊻ 16 ㊼ 49 ㊽ 64 ㊾ 20 ㊿ 24

Page 55

① 63 ② 35 ③ 32 ④ 48 ⑤ 20 ⑥ 35 ⑦ 48 ⑧ 81 ⑨ 16 ⑩ 35
⑪ 35 ⑫ 24 ⑬ 54 ⑭ 63 ⑮ 20 ⑯ 54 ⑰ 56 ⑱ 30 ⑲ 56 ⑳ 45
㉑ 42 ㉒ 64 ㉓ 45 ㉔ 30 ㉕ 54 ㉖ 35 ㉗ 28 ㉘ 28 ㉙ 40 ㉚ 54
㉛ 42 ㉜ 72 ㉝ 24 ㉞ 24 ㉟ 72 ㊱ 45 ㊲ 40 ㊳ 32 ㊴ 25 ㊵ 42
㊶ 40 ㊷ 35 ㊸ 64 ㊹ 36 ㊺ 49 ㊻ 36 ㊼ 20 ㊽ 63 ㊾ 72 ㊿ 36

Page 56

① 48 ② 25 ③ 81 ④ 28 ⑤ 32 ⑥ 28 ⑦ 24 ⑧ 30 ⑨ 48 ⑩ 63
⑪ 36 ⑫ 24 ⑬ 24 ⑭ 56 ⑮ 56 ⑯ 45 ⑰ 54 ⑱ 49 ⑲ 45 ⑳ 30
㉑ 35 ㉒ 63 ㉓ 48 ㉔ 40 ㉕ 20 ㉖ 54 ㉗ 40 ㉘ 16 ㉙ 72 ㉚ 42
㉛ 35 ㉜ 64 ㉝ 28 ㉞ 36 ㉟ 54 ㊱ 54 ㊲ 28 ㊳ 30 ㊴ 48 ㊵ 36
㊶ 36 ㊷ 36 ㊸ 35 ㊹ 36 ㊺ 20 ㊻ 64 ㊼ 42 ㊽ 45 ㊾ 32 ㊿ 72

Page 57

① 35 ② 24 ③ 40 ④ 81 ⑤ 32 ⑥ 24 ⑦ 35 ⑧ 30 ⑨ 64 ⑩ 28
⑪ 28 ⑫ 56 ⑬ 45 ⑭ 36 ⑮ 32 ⑯ 42 ⑰ 63 ⑱ 36 ⑲ 30 ⑳ 20
㉑ 54 ㉒ 45 ㉓ 24 ㉔ 28 ㉕ 45 ㉖ 28 ㉗ 48 ㉘ 56 ㉙ 36 ㉚ 42
㉛ 49 ㉜ 25 ㉝ 56 ㉞ 72 ㉟ 54 ㊱ 20 ㊲ 20 ㊳ 72 ㊴ 16 ㊵ 54
㊶ 36 ㊷ 28 ㊸ 45 ㊹ 40 ㊺ 42 ㊻ 63 ㊼ 32 ㊽ 36 ㊾ 25 ㊿ 48

Page 58

① 54 ② 45 ③ 40 ④ 28 ⑤ 28 ⑥ 24 ⑦ 32 ⑧ 81 ⑨ 56 ⑩ 56
⑪ 32 ⑫ 32 ⑬ 36 ⑭ 49 ⑮ 36 ⑯ 25 ⑰ 64 ⑱ 42 ⑲ 25 ⑳ 81
㉑ 35 ㉒ 30 ㉓ 56 ㉔ 30 ㉕ 36 ㉖ 45 ㉗ 63 ㉘ 16 ㉙ 48 ㉚ 48
㉛ 54 ㉜ 42 ㉝ 28 ㉞ 32 ㉟ 45 ㊱ 64 ㊲ 24 ㊳ 40 ㊴ 54 ㊵ 20
㊶ 20 ㊷ 32 ㊸ 54 ㊹ 72 ㊺ 20 ㊻ 42 ㊼ 63 ㊽ 72 ㊾ 24 ㊿ 35

Page 59

① 35　② 36　③ 63　④ 40　⑤ 24　⑥ 36　⑦ 54　⑧ 40　⑨ 42　⑩ 28
⑪ 28　⑫ 36　⑬ 25　⑭ 72　⑮ 28　⑯ 48　⑰ 56　⑱ 20　⑲ 81　⑳ 20
㉑ 45　㉒ 72　㉓ 32　㉔ 24　㉕ 25　㉖ 28　㉗ 54　㉘ 49　㉙ 32　㉚ 48
㉛ 42　㉜ 45　㉝ 36　㉞ 35　㉟ 64　㊱ 20　㊲ 36　㊳ 48　㊴ 16　㊵ 45
㊶ 32　㊷ 28　㊸ 72　㊹ 30　㊺ 63　㊻ 30　㊼ 28　㊽ 72　㊾ 45　㊿ 56

Page 60

① 63　② 16　③ 0　④ 0　⑤ 24　⑥ 42　⑦ 8　⑧ 54　⑨ 40　⑩ 15
⑪ 24　⑫ 27　⑬ 40　⑭ 36　⑮ 10　⑯ 0　⑰ 7　⑱ 12　⑲ 4　⑳ 9
㉑ 0　㉒ 35　㉓ 14　㉔ 4　㉕ 56　㉖ 5　㉗ 9　㉘ 32　㉙ 0　㉚ 0
㉛ 0　㉜ 2　㉝ 0　㉞ 72　㉟ 20　㊱ 0　㊲ 54　㊳ 5　㊴ 42　㊵ 6
㊶ 18　㊷ 8　㊸ 35　㊹ 15　㊺ 0　㊻ 32　㊼ 7　㊽ 16　㊾ 18　㊿ 12

Page 61

① 3　② 12　③ 6　④ 72　⑤ 12　⑥ 45　⑦ 45　⑧ 49　⑨ 8　⑩ 0
⑪ 25　⑫ 0　⑬ 63　⑭ 0　⑮ 18　⑯ 6　⑰ 24　⑱ 48　⑲ 1　⑳ 64
㉑ 2　㉒ 30　㉓ 36　㉔ 3　㉕ 6　㉖ 36　㉗ 0　㉘ 0　㉙ 56　㉚ 28
㉛ 0　㉜ 0　㉝ 8　㉞ 10　㉟ 24　㊱ 21　㊲ 16　㊳ 21　㊴ 81　㊵ 30
㊶ 48　㊷ 27　㊸ 20　㊹ 14　㊺ 9　㊻ 0　㊼ 0　㊽ 4　㊾ 18　㊿ 28

Page 62

① 12　② 16　③ 40　④ 0　⑤ 54　⑥ 15　⑦ 8　⑧ 63　⑨ 56　⑩ 3
⑪ 7　⑫ 42　⑬ 27　⑭ 27　⑮ 64　⑯ 45　⑰ 0　⑱ 24　⑲ 16　⑳ 0
㉑ 3　㉒ 40　㉓ 10　㉔ 0　㉕ 42　㉖ 0　㉗ 6　㉘ 36　㉙ 15　㉚ 7
㉛ 25　㉜ 0　㉝ 1　㉞ 49　㉟ 12　㊱ 9　㊲ 6　㊳ 0　㊴ 54　㊵ 12
㊶ 56　㊷ 0　㊸ 18　㊹ 48　㊺ 5　㊻ 28　㊼ 0　㊽ 14　㊾ 9　㊿ 24

Page 63

① 0　② 18　③ 0　④ 18　⑤ 4　⑥ 16　⑦ 21　⑧ 30　⑨ 32　⑩ 0
⑪ 24　⑫ 6　⑬ 35　⑭ 20　⑮ 4　⑯ 8　⑰ 36　⑱ 21　⑲ 0　⑳ 63
㉑ 0　㉒ 9　㉓ 8　㉔ 0　㉕ 0　㉖ 28　㉗ 8　㉘ 30　㉙ 45　㉚ 24
㉛ 4　㉜ 10　㉝ 0　㉞ 6　㉟ 12　㊱ 36　㊲ 72　㊳ 5　㊴ 14　㊵ 72
㊶ 81　㊷ 2　㊸ 32　㊹ 48　㊺ 0　㊻ 2　㊼ 35　㊽ 0　㊾ 18　㊿ 20

Page 64

① 0　② 2　③ 28　④ 18　⑤ 0　⑥ 32　⑦ 1　⑧ 81　⑨ 10　⑩ 18
⑪ 12　⑫ 63　⑬ 0　⑭ 36　⑮ 20　⑯ 63　⑰ 40　⑱ 14　⑲ 4　⑳ 0
㉑ 24　㉒ 40　㉓ 8　㉔ 10　㉕ 16　㉖ 9　㉗ 42　㉘ 0　㉙ 21　㉚ 54
㉛ 56　㉜ 12　㉝ 36　㉞ 18　㉟ 6　㊱ 0　㊲ 0　㊳ 25　㊴ 72　㊵ 7
㊶ 36　㊷ 0　㊸ 24　㊹ 35　㊺ 42　㊻ 4　㊼ 24　㊽ 2　㊾ 54　㊿ 16

Page 65

① 21　② 9　③ 20　④ 0　⑤ 6　⑥ 0　⑦ 27　⑧ 64　⑨ 8　⑩ 35
⑪ 15　⑫ 48　⑬ 72　⑭ 7　⑮ 0　⑯ 24　⑰ 18　⑱ 8　⑲ 15　⑳ 0
㉑ 5　㉒ 12　㉓ 0　㉔ 30　㉕ 27　㉖ 9　㉗ 30　㉘ 49　㉙ 0　㉚ 8
㉛ 12　㉜ 0　㉝ 6　㉞ 3　㉟ 4　㊱ 56　㊲ 0　㊳ 48　㊴ 0　㊵ 45
㊶ 0　㊷ 0　㊸ 5　㊹ 32　㊺ 45　㊻ 6　㊼ 14　㊽ 3　㊾ 28　㊿ 16

Page 66

① 0　② 3　③ 42　④ 12　⑤ 36　⑥ 8　⑦ 0　⑧ 24　⑨ 25　⑩ 36
⑪ 45　⑫ 27　⑬ 16　⑭ 64　⑮ 30　⑯ 56　⑰ 14　⑱ 9　⑲ 0　⑳ 0
㉑ 16　㉒ 5　㉓ 0　㉔ 7　㉕ 10　㉖ 24　㉗ 81　㉘ 12　㉙ 28　㉚ 6
㉛ 7　㉜ 36　㉝ 18　㉞ 15　㉟ 4　㊱ 8　㊲ 0　㊳ 35　㊴ 16　㊵ 0
㊶ 18　㊷ 28　㊸ 8　㊹ 45　㊺ 14　㊻ 5　㊼ 6　㊽ 0　㊾ 72　㊿ 48

Page 67

① 54　② 0　③ 56　④ 0　⑤ 27　⑥ 0　⑦ 12　⑧ 2　⑨ 6　⑩ 49
⑪ 20　⑫ 4　⑬ 3　⑭ 21　⑮ 0　⑯ 54　⑰ 8　⑱ 40　⑲ 42　⑳ 20
㉑ 0　㉒ 32　㉓ 0　㉔ 63　㉕ 48　㉖ 9　㉗ 10　㉘ 63　㉙ 4　㉚ 6
㉛ 2　㉜ 0　㉝ 30　㉞ 0　㉟ 24　㊱ 18　㊲ 24　㊳ 9　㊴ 0　㊵ 40
㊶ 32　㊷ 0　㊸ 15　㊹ 12　㊺ 21　㊻ 0　㊼ 35　㊽ 18　㊾ 72　㊿ 1

Page 68

① 36　② 56　③ 5　④ 5　⑤ 0　⑥ 56　⑦ 14　⑧ 0　⑨ 24　⑩ 27
⑪ 36　⑫ 0　⑬ 9　⑭ 42　⑮ 6　⑯ 30　⑰ 8　⑱ 0　⑲ 2　⑳ 9
㉑ 35　㉒ 48　㉓ 16　㉔ 16　㉕ 16　㉖ 3　㉗ 0　㉘ 21　㉙ 15　㉚ 0
㉛ 40　㉜ 1　㉝ 45　㉞ 0　㉟ 24　㊱ 15　㊲ 63　㊳ 12　㊴ 24　㊵ 0
㊶ 0　㊷ 6　㊸ 12　㊹ 0　㊺ 25　㊻ 32　㊼ 8　㊽ 8　㊾ 54　㊿ 7

Page 69
① 0 ② 12 ③ 14 ④ 27 ⑤ 63 ⑥ 54 ⑦ 0 ⑧ 32 ⑨ 8 ⑩ 10
⑪ 18 ⑫ 20 ⑬ 0 ⑭ 35 ⑮ 18 ⑯ 9 ⑰ 48 ⑱ 28 ⑲ 2 ⑳ 12
㉑ 4 ㉒ 21 ㉓ 72 ㉔ 18 ㉕ 6 ㉖ 0 ㉗ 30 ㉘ 0 ㉙ 24 ㉚ 28
㉛ 40 ㉜ 45 ㉝ 4 ㉞ 6 ㉟ 49 ㊱ 0 ㊲ 3 ㊳ 72 ㊴ 0 ㊵ 36
㊶ 18 ㊷ 81 ㊸ 42 ㊹ 4 ㊺ 0 ㊻ 10 ㊼ 20 ㊽ 7 ㊾ 0 ㊿ 64

Page 70
① 27 ② 48 ③ 12 ④ 42 ⑤ 35 ⑥ 24 ⑦ 24 ⑧ 16 ⑨ 16 ⑩ 45
⑪ 56 ⑫ 56 ⑬ 10 ⑭ 20 ⑮ 36 ⑯ 15 ⑰ 14 ⑱ 18 ⑲ 45 ⑳ 36
㉑ 8 ㉒ 9 ㉓ 18 ㉔ 48 ㉕ 27 ㉖ 12 ㉗ 40 ㉘ 16 ㉙ 49 ㉚ 40
㉛ 48 ㉜ 30 ㉝ 21 ㉞ 40 ㉟ 21 ㊱ 8 ㊲ 24 ㊳ 54 ㊴ 18 ㊵ 24
㊶ 4 ㊷ 28 ㊸ 81 ㊹ 30 ㊺ 18 ㊻ 45 ㊼ 21 ㊽ 14 ㊾ 32 ㊿ 63

Page 71
① 48 ② 63 ③ 32 ④ 6 ⑤ 14 ⑥ 14 ⑦ 45 ⑧ 30 ⑨ 15 ⑩ 8
⑪ 24 ⑫ 12 ⑬ 25 ⑭ 28 ⑮ 81 ⑯ 28 ⑰ 72 ⑱ 16 ⑲ 27 ⑳ 63
㉑ 20 ㉒ 72 ㉓ 32 ㉔ 6 ㉕ 15 ㉖ 54 ㉗ 16 ㉘ 35 ㉙ 32 ㉚ 42
㉛ 27 ㉜ 54 ㉝ 14 ㉞ 24 ㉟ 10 ㊱ 36 ㊲ 35 ㊳ 24 ㊴ 12 ㊵ 10
㊶ 30 ㊷ 28 ㊸ 18 ㊹ 54 ㊺ 64 ㊻ 18 ㊼ 6 ㊽ 24 ㊾ 10 ㊿ 12

Page 72
① 49 ② 30 ③ 20 ④ 20 ⑤ 18 ⑥ 21 ⑦ 48 ⑧ 48 ⑨ 9 ⑩ 18
⑪ 54 ⑫ 12 ⑬ 16 ⑭ 10 ⑮ 54 ⑯ 8 ⑰ 27 ⑱ 32 ⑲ 14 ⑳ 40
㉑ 18 ㉒ 28 ㉓ 21 ㉔ 30 ㉕ 49 ㉖ 64 ㉗ 12 ㉘ 9 ㉙ 36 ㉚ 6
㉛ 30 ㉜ 16 ㉝ 63 ㉞ 6 ㉟ 16 ㊱ 18 ㊲ 25 ㊳ 56 ㊴ 72 ㊵ 24
㊶ 24 ㊷ 24 ㊸ 63 ㊹ 16 ㊺ 35 ㊻ 18 ㊼ 16 ㊽ 27 ㊾ 15 ㊿ 28

Page 73
① 24 ② 81 ③ 36 ④ 12 ⑤ 32 ⑥ 27 ⑦ 14 ⑧ 10 ⑨ 14 ⑩ 40
⑪ 25 ⑫ 64 ⑬ 4 ⑭ 45 ⑮ 63 ⑯ 24 ⑰ 27 ⑱ 30 ⑲ 36 ⑳ 28
㉑ 12 ㉒ 42 ㉓ 15 ㉔ 56 ㉕ 8 ㉖ 45 ㉗ 30 ㉘ 8 ㉙ 36 ㉚ 72
㉛ 36 ㉜ 45 ㉝ 32 ㉞ 48 ㉟ 6 ㊱ 35 ㊲ 18 ㊳ 21 ㊴ 15 ㊵ 16
㊶ 16 ㊷ 45 ㊸ 35 ㊹ 56 ㊺ 18 ㊻ 32 ㊼ 56 ㊽ 24 ㊾ 6 ㊿ 42

Page 74
① 18 ② 20 ③ 12 ④ 40 ⑤ 14 ⑥ 42 ⑦ 18 ⑧ 24 ⑨ 35 ⑩ 72
⑪ 8 ⑫ 56 ⑬ 42 ⑭ 24 ⑮ 81 ⑯ 28 ⑰ 10 ⑱ 8 ⑲ 21 ⑳ 10
㉑ 45 ㉒ 24 ㉓ 15 ㉔ 20 ㉕ 18 ㉖ 12 ㉗ 32 ㉘ 35 ㉙ 16 ㉚ 49
㉛ 20 ㉜ 16 ㉝ 12 ㉞ 49 ㉟ 32 ㊱ 45 ㊲ 15 ㊳ 6 ㊴ 54 ㊵ 16
㊶ 30 ㊷ 72 ㊸ 27 ㊹ 16 ㊺ 30 ㊻ 72 ㊼ 32 ㊽ 10 ㊾ 21 ㊿ 24

Page 75
① 14 ② 18 ③ 36 ④ 20 ⑤ 48 ⑥ 10 ⑦ 21 ⑧ 35 ⑨ 48 ⑩ 18
⑪ 15 ⑫ 12 ⑬ 56 ⑭ 6 ⑮ 27 ⑯ 72 ⑰ 63 ⑱ 27 ⑲ 36 ⑳ 24
㉑ 63 ㉒ 12 ㉓ 21 ㉔ 36 ㉕ 28 ㉖ 45 ㉗ 27 ㉘ 64 ㉙ 36 ㉚ 4
㉛ 36 ㉜ 45 ㉝ 48 ㉞ 18 ㉟ 40 ㊱ 9 ㊲ 14 ㊳ 42 ㊴ 25 ㊵ 42
㊶ 16 ㊷ 6 ㊸ 30 ㊹ 6 ㊺ 28 ㊻ 8 ㊼ 36 ㊽ 16 ㊾ 40 ㊿ 54

Page 76
① 40 ② 28 ③ 36 ④ 21 ⑤ 16 ⑥ 25 ⑦ 12 ⑧ 12 ⑨ 30 ⑩ 63
⑪ 8 ⑫ 15 ⑬ 24 ⑭ 81 ⑮ 42 ⑯ 32 ⑰ 12 ⑱ 8 ⑲ 64 ⑳ 14
㉑ 36 ㉒ 20 ㉓ 48 ㉔ 28 ㉕ 40 ㉖ 6 ㉗ 36 ㉘ 30 ㉙ 6 ㉚ 40
㉛ 28 ㉜ 18 ㉝ 10 ㉞ 40 ㉟ 12 ㊱ 36 ㊲ 63 ㊳ 16 ㊴ 21 ㊵ 16
㊶ 18 ㊷ 18 ㊸ 56 ㊹ 18 ㊺ 35 ㊻ 63 ㊼ 12 ㊽ 12 ㊾ 45 ㊿ 24

Page 77
① 10 ② 14 ③ 24 ④ 24 ⑤ 9 ⑥ 12 ⑦ 64 ⑧ 56 ⑨ 45 ⑩ 27
⑪ 63 ⑫ 6 ⑬ 72 ⑭ 18 ⑮ 56 ⑯ 18 ⑰ 35 ⑱ 54 ⑲ 28 ⑳ 24
㉑ 48 ㉒ 32 ㉓ 45 ㉔ 15 ㉕ 32 ㉖ 49 ㉗ 54 ㉘ 27 ㉙ 24 ㉚ 4
㉛ 28 ㉜ 49 ㉝ 9 ㉞ 12 ㉟ 54 ㊱ 72 ㊲ 16 ㊳ 25 ㊴ 42 ㊵ 24
㊶ 18 ㊷ 18 ㊸ 35 ㊹ 16 ㊺ 20 ㊻ 8 ㊼ 15 ㊽ 16 ㊾ 54 ㊿ 30

Page 78
① 45 ② 18 ③ 64 ④ 12 ⑤ 42 ⑥ 36 ⑦ 18 ⑧ 42 ⑨ 20 ⑩ 6
⑪ 36 ⑫ 16 ⑬ 49 ⑭ 16 ⑮ 27 ⑯ 56 ⑰ 30 ⑱ 16 ⑲ 35 ⑳ 6
㉑ 45 ㉒ 72 ㉓ 25 ㉔ 18 ㉕ 45 ㉖ 14 ㉗ 12 ㉘ 20 ㉙ 24 ㉚ 28
㉛ 18 ㉜ 4 ㉝ 54 ㉞ 28 ㉟ 32 ㊱ 45 ㊲ 24 ㊳ 10 ㊴ 21 ㊵ 48
㊶ 35 ㊷ 36 ㊸ 15 ㊹ 4 ㊺ 27 ㊻ 6 ㊼ 32 ㊽ 30 ㊾ 32 ㊿ 20

Page 79

① 8 ② 18 ③ 54 ④ 40 ⑤ 28 ⑥ 30 ⑦ 35 ⑧ 21 ⑨ 18 ⑩ 56
⑪ 24 ⑫ 14 ⑬ 14 ⑭ 48 ⑮ 15 ⑯ 36 ⑰ 12 ⑱ 72 ⑲ 24 ⑳ 20
㉑ 63 ㉒ 30 ㉓ 32 ㉔ 63 ㉕ 56 ㉖ 9 ㉗ 72 ㉘ 8 ㉙ 54 ㉚ 12
㉛ 24 ㉜ 9 ㉝ 28 ㉞ 18 ㉟ 10 ㊱ 40 ㊲ 42 ㊳ 36 ㊴ 15 ㊵ 49
㊶ 4 ㊷ 48 ㊸ 27 ㊹ 10 ㊺ 24 ㊻ 16 ㊼ 63 ㊽ 48 ㊾ 10 ㊿ 81

Page 81

① 110 ② 70 ③ 50 ④ 90 ⑤ 60 ⑥ 100 ⑦ 120 ⑧ 10 ⑨ 40 ⑩ 30
⑪ 40 ⑫ 10 ⑬ 90 ⑭ 0 ⑮ 110 ⑯ 120 ⑰ 50 ⑱ 70 ⑲ 80 ⑳ 20
㉑ 60 ㉒ 90 ㉓ 100 ㉔ 80 ㉕ 30 ㉖ 20 ㉗ 110 ㉘ 0 ㉙ 50 ㉚ 40
㉛ 80 ㉜ 0 ㉝ 70 ㉞ 120 ㉟ 100 ㊱ 60 ㊲ 40 ㊳ 50 ㊴ 20 ㊵ 10
㊶ 70 ㊷ 30 ㊸ 120 ㊹ 20 ㊺ 10 ㊻ 90 ㊼ 80 ㊽ 100 ㊾ 110 ㊿ 50

Page 83

① 77 ② 121 ③ 44 ④ 99 ⑤ 11 ⑥ 66 ⑦ 55 ⑧ 88 ⑨ 33 ⑩ 132
⑪ 33 ⑫ 88 ⑬ 99 ⑭ 22 ⑮ 77 ⑯ 55 ⑰ 44 ⑱ 121 ⑲ 0 ⑳ 110
㉑ 11 ㉒ 99 ㉓ 66 ㉔ 0 ㉕ 132 ㉖ 110 ㉗ 77 ㉘ 22 ㉙ 44 ㉚ 33
㉛ 0 ㉜ 22 ㉝ 121 ㉞ 55 ㉟ 66 ㊱ 11 ㊲ 33 ㊳ 44 ㊴ 110 ㊵ 88
㊶ 121 ㊷ 132 ㊸ 55 ㊹ 110 ㊺ 88 ㊻ 99 ㊼ 0 ㊽ 66 ㊾ 77 ㊿ 44

Page 85

① 96 ② 120 ③ 12 ④ 144 ⑤ 0 ⑥ 24 ⑦ 132 ⑧ 60 ⑨ 72 ⑩ 36
⑪ 72 ⑫ 60 ⑬ 144 ⑭ 48 ⑮ 96 ⑯ 132 ⑰ 12 ⑱ 120 ⑲ 84 ⑳ 108
㉑ 0 ㉒ 144 ㉓ 24 ㉔ 84 ㉕ 36 ㉖ 108 ㉗ 96 ㉘ 48 ㉙ 12 ㉚ 72
㉛ 84 ㉜ 48 ㉝ 120 ㉞ 132 ㉟ 24 ㊱ 0 ㊲ 72 ㊳ 12 ㊴ 108 ㊵ 60
㊶ 120 ㊷ 36 ㊸ 132 ㊹ 108 ㊺ 60 ㊻ 144 ㊼ 84 ㊽ 24 ㊾ 96 ㊿ 12

Page 86

① 36 ② 0 ③ 40 ④ 33 ⑤ 18 ⑥ 0 ⑦ 24 ⑧ 20 ⑨ 32 ⑩ 12
⑪ 18 ⑫ 16 ⑬ 28 ⑭ 0 ⑮ 12 ⑯ 40 ⑰ 9 ⑱ 0 ⑲ 33 ⑳ 45
㉑ 11 ㉒ 50 ㉓ 0 ㉔ 18 ㉕ 0 ㉖ 45 ㉗ 24 ㉘ 60 ㉙ 7 ㉚ 24
㉛ 40 ㉜ 36 ㉝ 48 ㉞ 8 ㉟ 14 ㊱ 22 ㊲ 10 ㊳ 36 ㊴ 12 ㊵ 0
㊶ 24 ㊷ 10 ㊸ 18 ㊹ 30 ㊺ 27 ㊻ 8 ㊼ 0 ㊽ 32 ㊾ 55 ㊿ 21

Page 87

①　0　②44　③　6　④22　⑤　0　⑥24　⑦10　⑧35　⑨18　⑩30
⑪　7　⑫45　⑬24　⑭　0　⑮40　⑯30　⑰18　⑱44　⑲12　⑳　0
㉑50　㉒　0　㉓22　㉔　8　㉕60　㉖　7　㉗　0　㉘27　㉙24　㉚30
㉛33　㉜14　㉝45　㉞30　㉟　0　㊱36　㊲48　㊳　6　㊴60　㊵16
㊶40　㊷36　㊸　9　㊹28　㊺12　㊻21　㊼55　㊽16　㊾　0　㊿24

Page 88

①　0　②30　③24　④　7　⑤24　⑥30　⑦　0　⑧18　⑨　0　⑩40
⑪11　⑫16　⑬　0　⑭50　⑮40　⑯　0　⑰48　⑱35　⑲　7　⑳36
㉑28　㉒27　㉓45　㉔24　㉕55　㉖36　㉗　8　㉘30　㉙24　㉚　0
㉛24　㉜10　㉝　0　㉞32　㉟12　㊱14　㊲36　㊳10　㊴22　㊵60
㊶20　㊷36　㊸11　㊹33　㊺12　㊻32　㊼30　㊽　0　㊾21　㊿　6

Page 89

①55　②　0　③44　④14　⑤40　⑥20　⑦36　⑧18　⑨24　⑩　9
⑪24　⑫36　⑬　8　⑭45　⑮　0　⑯33　⑰24　⑱　0　⑲40　⑳35
㉑27　㉒55　㉓14　㉔32　㉕30　㉖24　㉗30　㉘12　㉙　0　㉚33
㉛　7　㉜12　㉝36　㉞　9　㉟35　㊱　0　㊲　0　㊳44　㊴30　㊵16
㊶　0　㊷10　㊸48　㊹　0　㊺22　㊻　6　㊼21　㊽16　㊾45　㊿　8

Page 90

①28　②16　③30　④　0　⑤35　⑥16　⑦48　⑧55　⑨40　⑩　6
⑪　0　⑫50　⑬32　⑭12　⑮　6　⑯44　⑰　7　⑱18　⑲　0　⑳21
㉑　9　㉒33　㉓22　㉔35　㉕24　㉖21　㉗　0　㉘18　㉙　8　㉚48
㉛30　㉜　0　㉝24　㉞10　㉟40　㊱45　㊲11　㊳　0　㊴60　㊵14
㊶30　㊷11　㊸　0　㊹36　㊺　0　㊻10　㊼16　㊽40　㊾27　㊿　0

Page 91

①24　②36　③12　④45　⑤20　⑥30　⑦11　⑧24　⑨35　⑩　0
⑪　8　⑫21　⑬　0　⑭22　⑮44　⑯36　⑰35　⑱36　⑲　6　⑳18
㉑33　㉒24　㉓45　㉔10　㉕18　㉖　8　㉗16　㉘　0　㉙48　㉚36
㉛　0　㉜40　㉝21　㉞　0　㉟18　㊱28　㊲24　㊳12　㊴18　㊵50
㊶44　㊷　0　㊸　7　㊹32　㊺60　㊻　0　㊼27　㊽50　㊾22　㊿　0

Page 92

① 30 ② 20 ③ 0 ④ 32 ⑤ 18 ⑥ 20 ⑦ 45 ⑧ 36 ⑨ 35 ⑩ 11

⑪ 36 ⑫ 21 ⑬ 50 ⑭ 22 ⑮ 11 ⑯ 60 ⑰ 6 ⑱ 16 ⑲ 32 ⑳ 0

㉑ 8 ㉒ 0 ㉓ 24 ㉔ 18 ㉕ 18 ㉖ 0 ㉗ 28 ㉘ 0 ㉙ 10 ㉚ 45

㉛ 0 ㉜ 44 ㉝ 55 ㉞ 7 ㉟ 30 ㊱ 24 ㊲ 12 ㊳ 44 ㊴ 27 ㊵ 12

㊶ 33 ㊷ 12 ㊸ 36 ㊹ 0 ㊺ 24 ㊻ 7 ㊼ 20 ㊽ 35 ㊾ 0 ㊿ 40

Page 93

① 18 ② 40 ③ 9 ④ 24 ⑤ 14 ⑥ 33 ⑦ 12 ⑧ 0 ⑨ 18 ⑩ 48

⑪ 10 ⑫ 0 ⑬ 28 ⑭ 24 ⑮ 60 ⑯ 0 ⑰ 18 ⑱ 40 ⑲ 11 ⑳ 16

㉑ 0 ㉒ 18 ㉓ 24 ㉔ 7 ㉕ 0 ㉖ 10 ㉗ 20 ㉘ 24 ㉙ 45 ㉚ 0

㉛ 32 ㉜ 30 ㉝ 0 ㉞ 48 ㉟ 16 ㊱ 30 ㊲ 55 ㊳ 9 ㊴ 0 ㊵ 21

㊶ 60 ㊷ 44 ㊸ 6 ㊹ 50 ㊺ 27 ㊻ 40 ㊼ 0 ㊽ 21 ㊾ 24 ㊿ 28

Page 94

① 7 ② 0 ③ 33 ④ 45 ⑤ 14 ⑥ 0 ⑦ 6 ⑧ 16 ⑨ 11 ⑩ 40

⑪ 30 ⑫ 22 ⑬ 12 ⑭ 0 ⑮ 40 ⑯ 8 ⑰ 28 ⑱ 0 ⑲ 45 ⑳ 21

㉑ 36 ㉒ 24 ㉓ 0 ㉔ 14 ㉕ 0 ㉖ 21 ㉗ 55 ㉘ 30 ㉙ 48 ㉚ 6

㉛ 33 ㉜ 50 ㉝ 10 ㉞ 44 ㉟ 24 ㊱ 18 ㊲ 32 ㊳ 50 ㊴ 12 ㊵ 0

㊶ 20 ㊷ 32 ㊸ 30 ㊹ 18 ㊺ 35 ㊻ 44 ㊼ 0 ㊽ 11 ㊾ 27 ㊿ 60

Page 95

① 0 ② 9 ③ 24 ④ 18 ⑤ 0 ⑥ 20 ⑦ 32 ⑧ 36 ⑨ 14 ⑩ 40

⑪ 48 ⑫ 21 ⑬ 55 ⑭ 0 ⑮ 8 ⑯ 18 ⑰ 14 ⑱ 9 ⑲ 40 ⑳ 0

㉑ 24 ㉒ 0 ㉓ 18 ㉔ 44 ㉕ 30 ㉖ 48 ㉗ 0 ㉘ 35 ㉙ 6 ㉚ 18

㉛ 45 ㉜ 24 ㉝ 21 ㉞ 40 ㉟ 0 ㊱ 7 ㊲ 10 ㊳ 24 ㊴ 30 ㊵ 22

㊶ 8 ㊷ 50 ㊸ 28 ㊹ 12 ㊺ 12 ㊻ 60 ㊼ 27 ㊽ 22 ㊾ 0 ㊿ 55

Page 96

① 63 ② 90 ③ 144 ④ 56 ⑤ 80 ⑥ 66 ⑦ 66 ⑧ 56 ⑨ 108 ⑩ 70

⑪ 110 ⑫ 88 ⑬ 60 ⑭ 132 ⑮ 70 ⑯ 108 ⑰ 72 ⑱ 72 ⑲ 63 ⑳ 66

㉑ 72 ㉒ 84 ㉓ 72 ㉔ 54 ㉕ 70 ㉖ 48 ㉗ 60 ㉘ 80 ㉙ 121 ㉚ 84

㉛ 64 ㉜ 63 ㉝ 100 ㉞ 99 ㉟ 144 ㊱ 88 ㊲ 120 ㊳ 36 ㊴ 56 ㊵ 81

㊶ 72 ㊷ 77 ㊸ 132 ㊹ 48 ㊺ 63 ㊻ 90 ㊼ 72 ㊽ 49 ㊾ 96 ㊿ 110

Page 97

① 80　② 54　③ 77　④ 120　⑤ 48　⑥ 120　⑦ 56　⑧ 99　⑨ 80　⑩ 121

⑪ 63　⑫ 96　⑬ 96　⑭ 90　⑮ 110　⑯ 49　⑰ 132　⑱ 63　⑲ 36　⑳ 54

㉑ 121　㉒ 80　㉓ 54　㉔ 88　㉕ 54　㉖ 108　㉗ 42　㉘ 42　㉙ 70　㉚ 96

㉛ 120　㉜ 96　㉝ 70　㉞ 42　㉟ 120　㊱ 88　㊲ 96　㊳ 121　㊴ 81　㊵ 42

㊶ 70　㊷ 60　㊸ 77　㊹ 99　㊺ 88　㊻ 42　㊼ 81　㊽ 72　㊾ 80　㊿ 96

Page 98

① 72　② 77　③ 108　④ 60　⑤ 90　⑥ 56　⑦ 88　⑧ 60　⑨ 144　⑩ 54

⑪ 99　⑫ 70　⑬ 56　⑭ 132　⑮ 42　⑯ 99　⑰ 120　⑱ 72　⑲ 66　⑳ 56

㉑ 72　㉒ 72　㉓ 120　㉔ 88　㉕ 54　㉖ 80　㉗ 72　㉘ 70　㉙ 77　㉚ 54

㉛ 100　㉜ 66　㉝ 63　㉞ 84　㉟ 108　㊱ 110　㊲ 84　㊳ 64　㊴ 60　㊵ 132

㊶ 96　㊷ 66　㊸ 63　㊹ 80　㊺ 66　㊻ 108　㊼ 96　㊽ 36　㊾ 90　㊿ 49

Page 99

① 90　② 96　③ 42　④ 81　⑤ 80　⑥ 84　⑦ 60　⑧ 121　⑨ 70　⑩ 77

⑪ 66　⑫ 90　⑬ 120　⑭ 77　⑮ 99　⑯ 36　⑰ 63　⑱ 72　⑲ 64　⑳ 96

㉑ 77　㉒ 90　㉓ 88　㉔ 110　㉕ 96　㉖ 99　㉗ 48　㉘ 48　㉙ 42　㉚ 120

㉛ 120　㉜ 90　㉝ 54　㉞ 48　㉟ 81　㊱ 110　㊲ 84　㊳ 77　㊴ 132　㊵ 48

㊶ 42　㊷ 72　㊸ 66　㊹ 84　㊺ 70　㊻ 48　㊼ 132　㊽ 96　㊾ 90　㊿ 90

Page 100

① 88　② 108　③ 70　④ 70　⑤ 66　⑥ 72　⑦ 72　⑧ 121　⑨ 80　⑩ 60

⑪ 54　⑫ 132　⑬ 96　⑭ 90　⑮ 132　⑯ 63　⑰ 56　⑱ 56　⑲ 90　⑳ 72

㉑ 56　㉒ 100　㉓ 56　㉔ 54　㉕ 60　㉖ 88　㉗ 36　㉘ 84　㉙ 108　㉚ 77

㉛ 77　㉜ 90　㉝ 72　㉞ 96　㉟ 70　㊱ 63　㊲ 120　㊳ 48　㊴ 121　㊵ 72

㊶ 60　㊷ 99　㊸ 84　㊹ 88　㊺ 90　㊻ 48　㊼ 60　㊽ 110　㊾ 49　㊿ 144

Page 101

① 66　② 64　③ 120　④ 42　⑤ 42　⑥ 120　⑦ 121　⑧ 81　⑨ 84　⑩ 108

⑪ 90　⑫ 49　⑬ 110　⑭ 108　⑮ 54　⑯ 110　⑰ 84　⑱ 88　⑲ 48　⑳ 64

㉑ 108　㉒ 66　㉓ 54　㉔ 63　㉕ 64　㉖ 63　㉗ 80　㉘ 66　㉙ 132　㉚ 110

㉛ 110　㉜ 49　㉝ 60　㉞ 66　㉟ 42　㊱ 63　㊲ 120　㊳ 108　㊴ 72　㊵ 80

㊶ 132　㊷ 36　㊸ 99　㊹ 96　㊺ 132　㊻ 80　㊼ 72　㊽ 60　㊾ 66　㊿ 49

Page 102

① 72 ② 121 ③ 80 ④ 49 ⑤ 60 ⑥ 108 ⑦ 72 ⑧ 120 ⑨ 60 ⑩ 42
⑪ 54 ⑫ 90 ⑬ 88 ⑭ 90 ⑮ 132 ⑯ 88 ⑰ 42 ⑱ 64 ⑲ 77 ⑳ 108
㉑ 64 ㉒ 70 ㉓ 42 ㉔ 132 ㉕ 42 ㉖ 80 ㉗ 72 ㉘ 77 ㉙ 81 ㉚ 96
㉛ 70 ㉜ 77 ㉝ 66 ㉞ 54 ㉟ 80 ㊱ 63 ㊲ 110 ㊳ 96 ㊴ 120 ㊵ 66
㊶ 120 ㊷ 108 ㊸ 72 ㊹ 80 ㊺ 77 ㊻ 36 ㊼ 120 ㊽ 84 ㊾ 56 ㊿ 99

Page 103

① 60 ② 48 ③ 63 ④ 48 ⑤ 84 ⑥ 110 ⑦ 120 ⑧ 99 ⑨ 77 ⑩ 81
⑪ 77 ⑫ 56 ⑬ 100 ⑭ 121 ⑮ 54 ⑯ 84 ⑰ 72 ⑱ 72 ⑲ 96 ⑳ 48
㉑ 81 ㉒ 60 ㉓ 132 ㉔ 63 ㉕ 48 ㉖ 88 ㉗ 56 ㉘ 144 ㉙ 132 ㉚ 100
㉛ 100 ㉜ 56 ㉝ 42 ㉞ 144 ㉟ 48 ㊱ 63 ㊲ 110 ㊳ 81 ㊴ 66 ㊵ 56
㊶ 132 ㊷ 72 ㊸ 108 ㊹ 54 ㊺ 90 ㊻ 56 ㊼ 66 ㊽ 120 ㊾ 60 ㊿ 56

Page 104

① 121 ② 100 ③ 84 ④ 81 ⑤ 84 ⑥ 48 ⑦ 48 ⑧ 132 ⑨ 132 ⑩ 63
⑪ 56 ⑫ 96 ⑬ 60 ⑭ 96 ⑮ 110 ⑯ 70 ⑰ 99 ⑱ 42 ⑲ 90 ⑳ 48
㉑ 42 ㉒ 108 ㉓ 99 ㉔ 60 ㉕ 63 ㉖ 72 ㉗ 42 ㉘ 90 ㉙ 64 ㉚ 77
㉛ 108 ㉜ 90 ㉝ 70 ㉞ 88 ㉟ 84 ㊱ 72 ㊲ 120 ㊳ 36 ㊴ 132 ㊵ 110
㊶ 72 ㊷ 88 ㊸ 56 ㊹ 72 ㊺ 90 ㊻ 77 ㊼ 72 ㊽ 99 ㊾ 63 ㊿ 80

Page 105

① 84 ② 66 ③ 72 ④ 49 ⑤ 54 ⑥ 120 ⑦ 132 ⑧ 80 ⑨ 90 ⑩ 64
⑪ 90 ⑫ 63 ⑬ 144 ⑭ 100 ⑮ 56 ⑯ 99 ⑰ 56 ⑱ 121 ⑲ 36 ⑳ 66
㉑ 64 ㉒ 84 ㉓ 60 ㉔ 72 ㉕ 66 ㉖ 70 ㉗ 54 ㉘ 66 ㉙ 110 ㉚ 144
㉛ 144 ㉜ 63 ㉝ 63 ㉞ 66 ㉟ 49 ㊱ 72 ㊲ 120 ㊳ 64 ㊴ 110 ㊵ 54
㊶ 110 ㊷ 42 ㊸ 88 ㊹ 88 ㊺ 96 ㊻ 54 ㊼ 110 ㊽ 72 ㊾ 84 ㊿ 63

Page 106

① 4 ② 81 ③ 60 ④ 21 ⑤ 24 ⑥ 0 ⑦ 3 ⑧ 0 ⑨ 110 ⑩ 24
⑪ 0 ⑫ 30 ⑬ 11 ⑭ 96 ⑮ 18 ⑯ 0 ⑰ 77 ⑱ 3 ⑲ 30 ⑳ 108
㉑ 45 ㉒ 12 ㉓ 8 ㉔ 40 ㉕ 84 ㉖ 108 ㉗ 0 ㉘ 12 ㉙ 80 ㉚ 22
㉛ 64 ㉜ 50 ㉝ 42 ㉞ 0 ㉟ 33 ㊱ 10 ㊲ 20 ㊳ 48 ㊴ 0 ㊵ 77
㊶ 4 ㊷ 22 ㊸ 72 ㊹ 4 ㊺ 72 ㊻ 0 ㊼ 72 ㊽ 63 ㊾ 20 ㊿ 9

Page 107

① 6 ② 12 ③ 0 ④ 0 ⑤ 15 ⑥ 24 ⑦ 36 ⑧ 120 ⑨ 55 ⑩ 56
⑪ 84 ⑫ 35 ⑬ 27 ⑭ 0 ⑮ 80 ⑯ 40 ⑰ 0 ⑱ 48 ⑲ 66 ⑳ 1
㉑ 27 ㉒ 88 ㉓ 40 ㉔ 88 ㉕ 2 ㉖ 0 ㉗ 7 ㉘ 25 ㉙ 6 ㉚ 60
㉛ 66 ㉜ 48 ㉝ 20 ㉞ 9 ㉟ 70 ㊱ 21 ㊲ 36 ㊳ 24 ㊴ 16 ㊵ 0
㊶ 0 ㊷ 12 ㊸ 6 ㊹ 60 ㊺ 0 ㊻ 44 ㊼ 100 ㊽ 99 ㊾ 49 ㊿ 45

Page 108

① 72 ② 6 ③ 9 ④ 24 ⑤ 20 ⑥ 20 ⑦ 0 ⑧ 110 ⑨ 12 ⑩ 33
⑪ 40 ⑫ 56 ⑬ 0 ⑭ 5 ⑮ 8 ⑯ 110 ⑰ 4 ⑱ 144 ⑲ 0 ⑳ 6
㉑ 0 ㉒ 48 ㉓ 132 ㉔ 18 ㉕ 4 ㉖ 1 ㉗ 90 ㉘ 88 ㉙ 35 ㉚ 18
㉛ 55 ㉜ 54 ㉝ 10 ㉞ 90 ㉟ 16 ㊱ 84 ㊲ 0 ㊳ 48 ㊴ 0 ㊵ 8
㊶ 36 ㊷ 16 ㊸ 11 ㊹ 0 ㊺ 30 ㊻ 20 ㊼ 5 ㊽ 4 ㊾ 63 ㊿ 96

Page 109

① 0 ② 45 ③ 20 ④ 10 ⑤ 0 ⑥ 8 ⑦ 24 ⑧ 6 ⑨ 18 ⑩ 22
⑪ 2 ⑫ 0 ⑬ 48 ⑭ 70 ⑮ 66 ⑯ 45 ⑰ 10 ⑱ 3 ⑲ 22 ⑳ 0
㉑ 12 ㉒ 10 ㉓ 28 ㉔ 22 ㉕ 0 ㉖ 30 ㉗ 48 ㉘ 0 ㉙ 64 ㉚ 30
㉛ 10 ㉜ 121 ㉝ 48 ㉞ 72 ㉟ 14 ㊱ 32 ㊲ 12 ㊳ 9 ㊴ 12 ㊵ 50
㊶ 80 ㊷ 0 ㊸ 108 ㊹ 0 ㊺ 80 ㊻ 8 ㊼ 42 ㊽ 12 ㊾ 8 ㊿ 9

Page 110

① 40 ② 0 ③ 0 ④ 9 ⑤ 88 ⑥ 10 ⑦ 30 ⑧ 45 ⑨ 16 ⑩ 70
⑪ 5 ⑫ 55 ⑬ 12 ⑭ 0 ⑮ 0 ⑯ 50 ⑰ 6 ⑱ 27 ⑲ 96 ⑳ 0
㉑ 48 ㉒ 33 ㉓ 81 ㉔ 56 ㉕ 0 ㉖ 0 ㉗ 60 ㉘ 100 ㉙ 77 ㉚ 8
㉛ 63 ㉜ 96 ㉝ 24 ㉞ 20 ㉟ 10 ㊱ 99 ㊲ 84 ㊳ 132 ㊴ 30 ㊵ 2
㊶ 63 ㊷ 20 ㊸ 0 ㊹ 66 ㊺ 28 ㊻ 15 ㊼ 0 ㊽ 0 ㊾ 44 ㊿ 15

Page 111

① 60 ② 32 ③ 10 ④ 0 ⑤ 36 ⑥ 0 ⑦ 44 ⑧ 0 ⑨ 24 ⑩ 27
⑪ 0 ⑫ 12 ⑬ 20 ⑭ 55 ⑮ 72 ⑯ 84 ⑰ 0 ⑱ 0 ⑲ 20 ⑳ 54
㉑ 0 ㉒ 14 ㉓ 121 ㉔ 18 ㉕ 24 ㉖ 35 ㉗ 9 ㉘ 144 ㉙ 50 ㉚ 64
㉛ 16 ㉜ 90 ㉝ 80 ㉞ 36 ㉟ 33 ㊱ 5 ㊲ 0 ㊳ 0 ㊴ 77 ㊵ 35
㊶ 25 ㊷ 0 ㊸ 12 ㊹ 0 ㊺ 50 ㊻ 22 ㊼ 88 ㊽ 8 ㊾ 3 ㊿ 0

Page 112

① 60 ② 132 ③ 8 ④ 18 ⑤ 110 ⑥ 0 ⑦ 7 ⑧ 0 ⑨ 0 ⑩ 20
⑪ 0 ⑫ 3 ⑬ 42 ⑭ 6 ⑮ 55 ⑯ 0 ⑰ 54 ⑱ 4 ⑲ 44 ⑳ 12
㉑ 48 ㉒ 20 ㉓ 18 ㉔ 0 ㉕ 10 ㉖ 11 ㉗ 0 ㉘ 25 ㉙ 18 ㉚ 72
㉛ 54 ㉜ 0 ㉝ 99 ㉞ 0 ㉟ 6 ㊱ 6 ㊲ 16 ㊳ 120 ㊴ 0 ㊵ 60
㊶ 8 ㊷ 30 ㊸ 5 ㊹ 70 ㊺ 72 ㊻ 0 ㊼ 11 ㊽ 110 ㊾ 36 ㊿ 2

Page 113

① 35 ② 132 ③ 0 ④ 0 ⑤ 8 ⑥ 5 ⑦ 120 ⑧ 0 ⑨ 48 ⑩ 81
⑪ 9 ⑫ 40 ⑬ 12 ⑭ 0 ⑮ 0 ⑯ 48 ⑰ 0 ⑱ 4 ⑲ 30 ⑳ 14
㉑ 22 ㉒ 36 ㉓ 30 ㉔ 54 ㉕ 84 ㉖ 0 ㉗ 20 ㉘ 32 ㉙ 5 ㉚ 0
㉛ 66 ㉜ 45 ㉝ 0 ㉞ 24 ㉟ 27 ㊱ 10 ㊲ 2 ㊳ 12 ㊴ 40 ㊵ 0
㊶ 0 ㊷ 7 ㊸ 24 ㊹ 4 ㊺ 0 ㊻ 60 ㊼ 0 ㊽ 72 ㊾ 90 ㊿ 88

Page 114

① 6 ② 20 ③ 16 ④ 18 ⑤ 21 ⑥ 0 ⑦ 60 ⑧ 35 ⑨ 0 ⑩ 144
⑪ 55 ⑫ 6 ⑬ 0 ⑭ 18 ⑮ 30 ⑯ 60 ⑰ 0 ⑱ 72 ⑲ 30 ⑳ 8
㉑ 40 ㉒ 63 ㉓ 56 ㉔ 0 ㉕ 22 ㉖ 10 ㉗ 40 ㉘ 72 ㉙ 9 ㉚ 0
㉛ 63 ㉜ 0 ㉝ 6 ㉞ 5 ㉟ 0 ㊱ 8 ㊲ 120 ㊳ 84 ㊴ 50 ㊵ 0
㊶ 96 ㊷ 0 ㊸ 24 ㊹ 70 ㊺ 36 ㊻ 10 ㊼ 6 ㊽ 55 ㊾ 1 ㊿ 54

Page 115

① 120 ② 3 ③ 0 ④ 25 ⑤ 90 ⑥ 12 ⑦ 28 ⑧ 0 ⑨ 0 ⑩ 14
⑪ 4 ⑫ 110 ⑬ 24 ⑭ 5 ⑮ 0 ⑯ 72 ⑰ 25 ⑱ 24 ⑲ 0 ⑳ 80
㉑ 45 ㉒ 0 ㉓ 7 ㉔ 0 ㉕ 10 ㉖ 60 ㉗ 88 ㉘ 80 ㉙ 36 ㉚ 0
㉛ 0 ㉜ 84 ㉝ 0 ㉞ 32 ㉟ 2 ㊱ 66 ㊲ 18 ㊳ 2 ㊴ 84 ㊵ 45
㊶ 30 ㊷ 20 ㊸ 9 ㊹ 20 ㊺ 30 ㊻ 0 ㊼ 0 ㊽ 0 ㊾ 22 ㊿ 40

Page 116

① 9 ② 132 ③ 28 ④ 72 ⑤ 35 ⑥ 110 ⑦ 96 ⑧ 8 ⑨ 96 ⑩ 24
⑪ 108 ⑫ 36 ⑬ 16 ⑭ 8 ⑮ 55 ⑯ 54 ⑰ 99 ⑱ 84 ⑲ 30 ⑳ 70
㉑ 27 ㉒ 50 ㉓ 36 ㉔ 66 ㉕ 60 ㉖ 56 ㉗ 20 ㉘ 44 ㉙ 36 ㉚ 80
㉛ 18 ㉜ 120 ㉝ 50 ㉞ 56 ㉟ 45 ㊱ 16 ㊲ 33 ㊳ 48 ㊴ 63 ㊵ 15
㊶ 20 ㊷ 90 ㊸ 132 ㊹ 35 ㊺ 18 ㊻ 99 ㊼ 60 ㊽ 64 ㊾ 12 ㊿ 40

Page 117

① 15　② 12　③ 16　④ 12　⑤ 24　⑥ 48　⑦ 100　⑧ 33　⑨ 36　⑩ 88
⑪ 20　⑫ 21　⑬ 144　⑭ 55　⑮ 42　⑯ 4　⑰ 77　⑱ 40　⑲ 22　⑳ 18
㉑ 60　㉒ 88　㉓ 66　㉔ 30　㉕ 42　㉖ 30　㉗ 48　㉘ 10　㉙ 72　㉚ 14
㉛ 77　㉜ 6　㉝ 44　㉞ 72　㉟ 24　㊱ 45　㊲ 30　㊳ 108　㊴ 6　㊵ 70
㊶ 49　㊷ 120　㊸ 36　㊹ 12　㊺ 32　㊻ 81　㊼ 84　㊽ 10　㊾ 72　㊿ 40

Page 118

① 27　② 66　③ 80　④ 35　⑤ 80　⑥ 63　⑦ 44　⑧ 6　⑨ 60　⑩ 110
⑪ 24　⑫ 4　⑬ 12　⑭ 50　⑮ 24　⑯ 22　⑰ 49　⑱ 88　⑲ 120　⑳ 72
㉑ 63　㉒ 60　㉓ 33　㉔ 77　㉕ 108　㉖ 40　㉗ 30　㉘ 60　㉙ 108　㉚ 24
㉛ 10　㉜ 72　㉝ 36　㉞ 32　㉟ 28　㊱ 20　㊲ 54　㊳ 55　㊴ 16　㊵ 12
㊶ 27　㊷ 42　㊸ 84　㊹ 32　㊺ 36　㊻ 12　㊼ 110　㊽ 20　㊾ 60　㊿ 12

Page 119

① 90　② 33　③ 25　④ 18　⑤ 36　⑥ 48　⑦ 54　⑧ 21　⑨ 70　⑩ 30
⑪ 8　⑫ 24　⑬ 132　⑭ 70　⑮ 88　⑯ 15　⑰ 56　⑱ 90　⑲ 18　⑳ 21
㉑ 48　㉒ 28　㉓ 72　㉔ 81　㉕ 96　㉖ 18　㉗ 22　㉘ 30　㉙ 121　㉚ 40
㉛ 48　㉜ 45　㉝ 14　㉞ 144　㉟ 36　㊱ 20　㊲ 44　㊳ 77　㊴ 9　㊵ 48
㊶ 64　㊷ 99　㊸ 132　㊹ 30　㊺ 10　㊻ 14　㊼ 96　㊽ 20　㊾ 8　㊿ 50

Page 120

① 40　② 22　③ 18　④ 42　⑤ 48　⑥ 99　⑦ 88　⑧ 28　⑨ 84　⑩ 60
⑪ 24　⑫ 14　⑬ 32　⑭ 54　⑮ 10　⑯ 20　⑰ 54　⑱ 33　⑲ 12　⑳ 132
㉑ 48　㉒ 40　㉓ 55　㉔ 90　㉕ 33　㉖ 21　㉗ 90　㉘ 12　㉙ 96　㉚ 80
㉛ 18　㉜ 120　㉝ 55　㉞ 96　㉟ 30　㊱ 42　㊲ 16　㊳ 70　㊴ 6　㊵ 25
㊶ 44　㊷ 60　㊸ 108　㊹ 15　㊺ 15　㊻ 4　㊼ 44　㊽ 56　㊾ 27　㊿ 70

Page 121

① 32　② 40　③ 63　④ 56　⑤ 48　⑥ 24　⑦ 110　⑧ 45　⑨ 36　⑩ 14
⑪ 35　⑫ 60　⑬ 132　⑭ 36　⑮ 120　⑯ 36　⑰ 27　⑱ 66　⑲ 8　⑳ 24
㉑ 60　㉒ 72　㉓ 6　㉔ 88　㉕ 9　㉖ 50　㉗ 77　㉘ 16　㉙ 110　㉚ 108
㉛ 24　㉜ 72　㉝ 63　㉞ 36　㉟ 64　㊱ 8　㊲ 50　㊳ 66　㊴ 20　㊵ 30
㊶ 36　㊷ 121　㊸ 30　㊹ 30　㊺ 49　㊻ 12　㊼ 144　㊽ 45　㊾ 16　㊿ 28

Page 122

① 25	② 6	③ 8	④ 36	⑤ 120	⑥ 77	⑦ 14	⑧ 72	⑨ 48	⑩ 16
⑪ 88	⑫ 99	⑬ 56	⑭ 48	⑮ 30	⑯ 44	⑰ 42	⑱ 4	⑲ 90	⑳ 132
㉑ 30	㉒ 30	㉓ 10	㉔ 28	㉕ 99	㉖ 12	㉗ 36	㉘ 12	㉙ 40	㉚ 21
㉛ 132	㉜ 24	㉝ 110	㉞ 84	㉟ 60	㊱ 36	㊲ 15	㊳ 24	㊴ 22	㊵ 50
㊶ 88	㊷ 18	㊸ 56	㊹ 20	㊺ 45	㊻ 33	㊼ 20	㊽ 42	㊾ 108	㊿ 27

Page 123

① 50	② 32	③ 72	④ 45	⑤ 64	⑥ 63	⑦ 33	⑧ 35	⑨ 24	⑩ 18
⑪ 90	⑫ 60	⑬ 16	⑭ 70	⑮ 48	⑯ 96	⑰ 14	⑱ 44	⑲ 24	⑳ 48
㉑ 80	㉒ 49	㉓ 27	㉔ 55	㉕ 18	㉖ 15	㉗ 18	㉘ 80	㉙ 8	㉚ 144
㉛ 36	㉜ 60	㉝ 63	㉞ 72	㉟ 35	㊱ 110	㊲ 40	㊳ 12	㊴ 40	㊵ 6
㊶ 24	㊷ 22	㊸ 36	㊹ 20	㊺ 54	㊻ 66	㊼ 96	㊽ 120	㊾ 77	㊿ 60

Page 124

① 36	② 24	③ 60	④ 84	⑤ 28	⑥ 33	⑦ 30	⑧ 90	⑨ 77	⑩ 24
⑪ 88	⑫ 72	⑬ 100	⑭ 48	⑮ 72	⑯ 16	⑰ 132	⑱ 15	⑲ 14	⑳ 12
㉑ 72	㉒ 35	㉓ 18	㉔ 22	㉕ 6	㉖ 35	㉗ 20	㉘ 96	㉙ 66	㉚ 50
㉛ 32	㉜ 55	㉝ 27	㉞ 40	㉟ 108	㊱ 108	㊲ 48	㊳ 14	㊴ 40	㊵ 54
㊶ 30	㊷ 60	㊸ 121	㊹ 45	㊺ 12	㊻ 64	㊼ 21	㊽ 70	㊾ 8	㊿ 45

Page 125

① 42	② 20	③ 28	④ 54	⑤ 110	⑥ 20	⑦ 15	⑧ 66	⑨ 144	⑩ 56
⑪ 81	⑫ 24	⑬ 33	⑭ 77	⑮ 8	⑯ 40	⑰ 55	⑱ 36	⑲ 80	⑳ 120
㉑ 99	㉒ 110	㉓ 16	㉔ 18	㉕ 10	㉖ 30	㉗ 27	㉘ 70	㉙ 6	㉚ 16
㉛ 32	㉜ 24	㉝ 99	㉞ 22	㉟ 60	㊱ 56	㊲ 18	㊳ 36	㊴ 60	㊵ 25
㊶ 20	㊷ 9	㊸ 4	㊹ 72	㊺ 63	㊻ 96	㊼ 44	㊽ 36	㊾ 80	㊿ 49

ABOUT THE AUTHOR

Dr. Chris McMullen has over 20 years of experience teaching university physics in California, Oklahoma, Pennsylvania, and Louisiana. Dr. McMullen is also an author of math and science workbooks. Whether in the classroom or as a writer, Dr. McMullen loves sharing knowledge and the art of motivating and engaging students.

The author earned his Ph.D. in phenomenological high-energy physics (particle physics) from Oklahoma State University in 2002. Originally from California, Chris McMullen earned his Master's degree from California State University, Northridge, where his thesis was in the field of electron spin resonance.

As a physics teacher, Dr. McMullen observed that many students lack fluency in fundamental math skills. In an effort to help students of all ages and levels master basic math skills, he published a series of math workbooks on arithmetic, fractions, long division, algebra, trigonometry, and calculus entitled *Improve Your Math Fluency*. Dr. McMullen has also published a variety of science books, including introductions to basic astronomy and chemistry concepts in addition to physics workbooks.

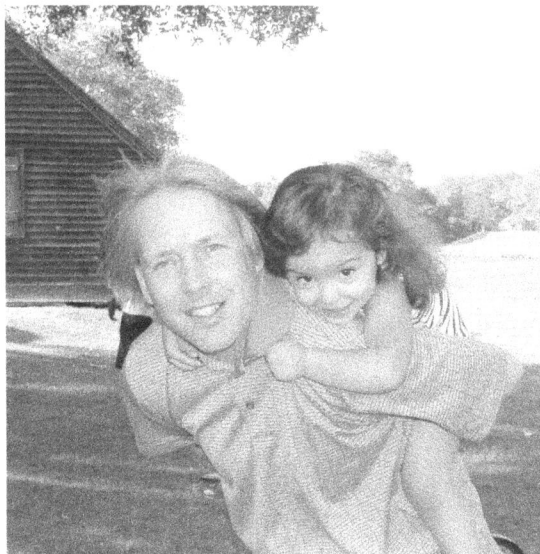

Author, Chris McMullen, Ph.D.

ARITHMETIC

For students who could benefit from additional arithmetic practice:
- Addition, subtraction, multiplication, and division facts
- Multi-digit addition and subtraction
- Addition and subtraction applied to clocks
- Multiplication with 10-20
- Multi-digit multiplication
- Long division with remainders
- Fractions
- Mixed fractions
- Decimals
- Fractions, decimals, and percentages

www.improveyourmathfluency.com

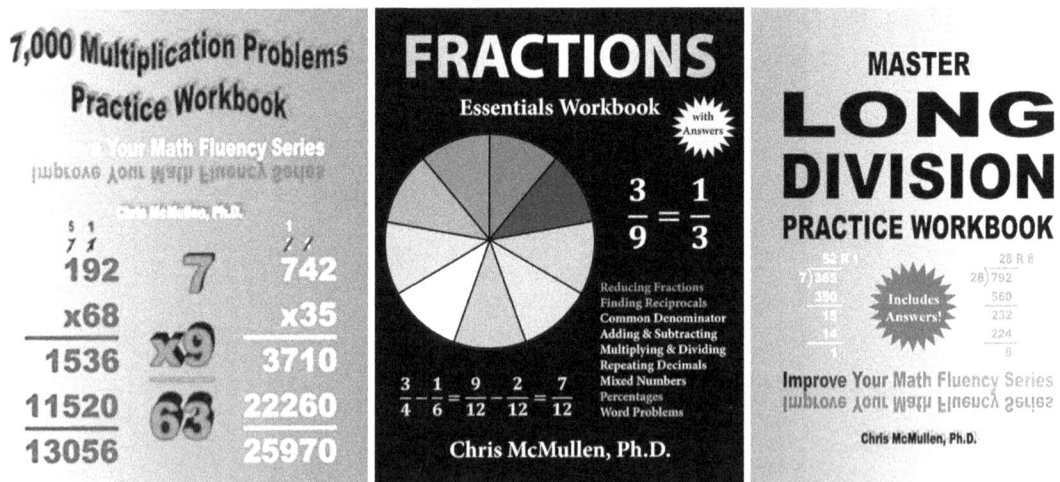

MATH

This series of math workbooks is geared toward practicing essential math skills:

- Algebra and trigonometry
- Geometry
- Calculus
- Fractions, decimals, and percentages
- Long division
- Multiplication and division
- Addition and subtraction

www.improveyourmathfluency.com

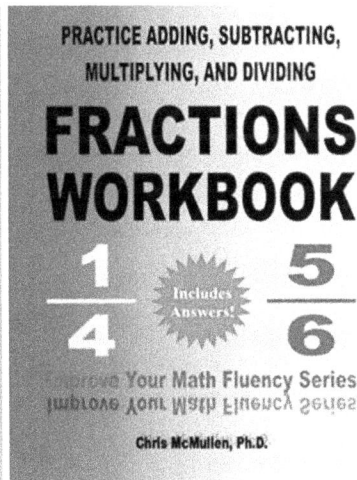

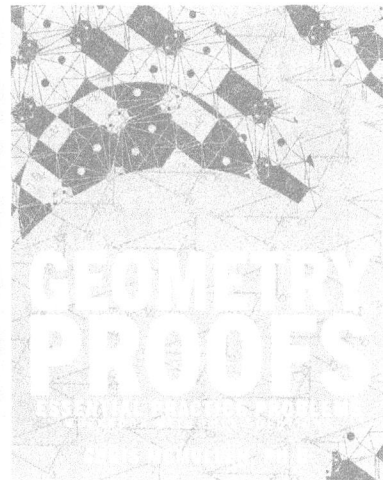

ALGEBRA

For students who need to improve their algebra skills:

- Isolating the unknown
- Quadratic equations
- Factoring
- Cross multiplying
- Systems of equations
- Straight line graphs
- Word problems

www.improveyourmathfluency.com

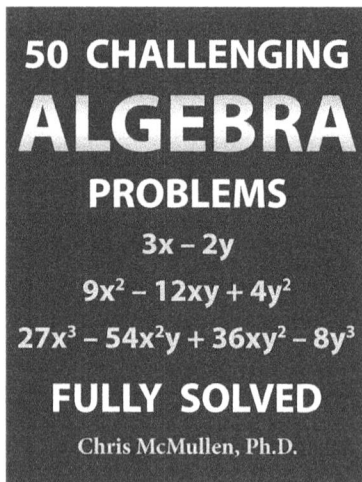

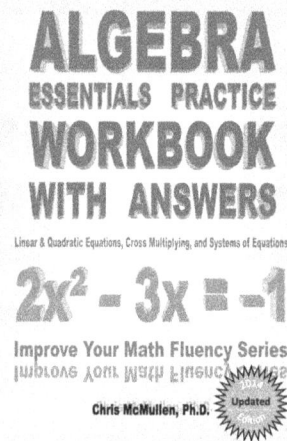

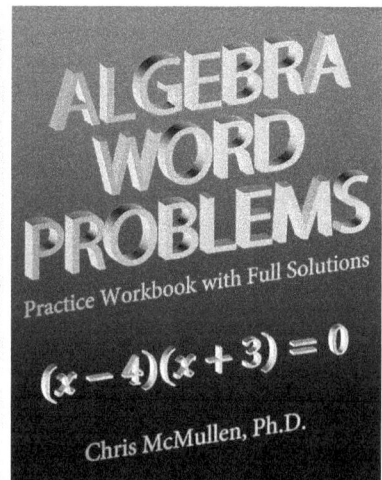

PUZZLES

The author of this book, Chris McMullen, enjoys solving puzzles. His favorite puzzle is Kakuro (kind of like a cross between crossword puzzles and Sudoku). He once taught a three-week summer course on puzzles. If you enjoy mathematical pattern puzzles, you might appreciate:

300+ Mathematical Pattern Puzzles

Number Pattern Recognition & Reasoning
- Pattern recognition
- Visual discrimination
- Analytical skills
- Logic and reasoning
- Analogies
- Mathematics

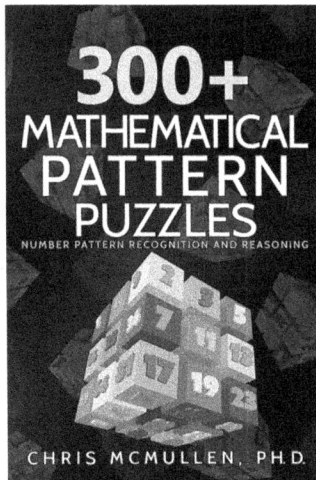

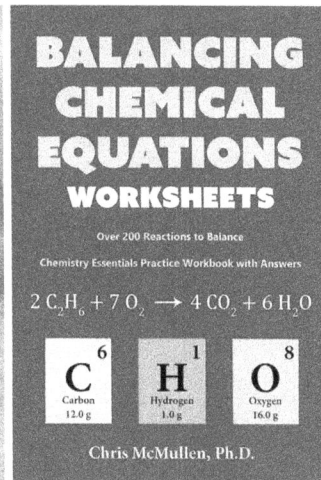

SCIENCE

Dr. McMullen has published a variety of **science** books, including:

- Basic astronomy concepts
- Basic chemistry concepts
- Balancing chemical reactions
- Calculus-based physics textbooks
- Calculus-based physics workbooks
- Calculus-based physics examples
- Trig-based physics workbooks
- Trig-based physics examples
- Creative physics problems
- Modern physics

www.monkeyphysicsblog.wordpress.com

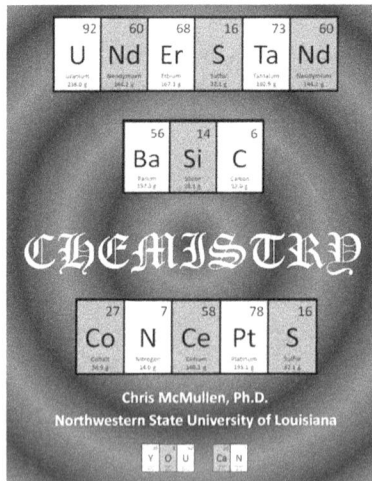

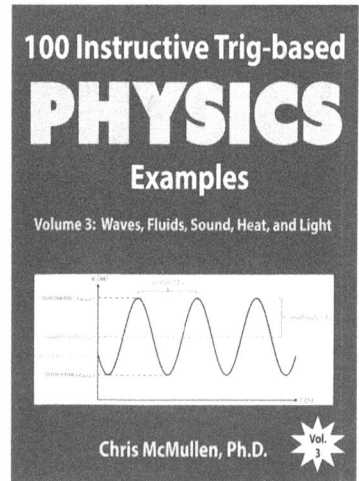

www.ingramcontent.com/pod-product-compliance
Lightning Source LLC
Chambersburg PA
CBHW081513040426
42447CB00013B/3211